K. Suresh Kumar
P.V. Subba Rao

Cultivation and Biochemical Constituents of Kappaphycus alvarez*ii*

K. Suresh Kumar
P.V. Subba Rao

Cultivation and Biochemical Constituents of Kappaphycus alvarezii

Cultivation of K. alvarezii

LAP LAMBERT Academic Publishing

Impressum / Imprint

Bibliografische Information der Deutschen Nationalbibliothek: Die Deutsche Nationalbibliothek verzeichnet diese Publikation in der Deutschen Nationalbibliografie; detaillierte bibliografische Daten sind im Internet über http://dnb.d-nb.de abrufbar.
Alle in diesem Buch genannten Marken und Produktnamen unterliegen warenzeichen-, marken- oder patentrechtlichem Schutz bzw. sind Warenzeichen oder eingetragene Warenzeichen der jeweiligen Inhaber. Die Wiedergabe von Marken, Produktnamen, Gebrauchsnamen, Handelsnamen, Warenbezeichnungen u.s.w. in diesem Werk berechtigt auch ohne besondere Kennzeichnung nicht zu der Annahme, dass solche Namen im Sinne der Warenzeichen- und Markenschutzgesetzgebung als frei zu betrachten wären und daher von jedermann benutzt werden dürften.

Bibliographic information published by the Deutsche Nationalbibliothek: The Deutsche Nationalbibliothek lists this publication in the Deutsche Nationalbibliografie; detailed bibliographic data are available in the Internet at http://dnb.d-nb.de.
Any brand names and product names mentioned in this book are subject to trademark, brand or patent protection and are trademarks or registered trademarks of their respective holders. The use of brand names, product names, common names, trade names, product descriptions etc. even without a particular marking in this work is in no way to be construed to mean that such names may be regarded as unrestricted in respect of trademark and brand protection legislation and could thus be used by anyone.

Coverbild / Cover image: www.ingimage.com

Verlag / Publisher:
LAP LAMBERT Academic Publishing
ist ein Imprint der / is a trademark of
OmniScriptum GmbH & Co. KG
Heinrich-Böcking-Str. 6-8, 66121 Saarbrücken, Deutschland / Germany
Email: info@lap-publishing.com

Herstellung: siehe letzte Seite /
Printed at: see last page
ISBN: 978-3-659-34302-5

Zugl. / Approved by: Bhavnagar, Bhavnagar University/CSMCRI-CSIR, 2008

K. Suresh Kumar
P.V. Subba Rao

Studies on Cultivation and the Biochemical Constituents of
Kappaphycus alvarezii (Doty) Doty

Dedicated to
My beloved Parents, Wife and Son

Acknowledgements

I seize this opportunity to convey my prodigious and profound gratitude to my guide Dr. P V Subba Rao, Scientist E II, whose exuberant leadership and beacon-like guidance stimulated me to accomplish my cherished ambition throughout the course of my doctoral program. With his finesse and farsightedness, he has astutely and adroitly guided me into this research meadow.

I wish to express my titanic thanks to our esteemed Director Dr. P K Ghosh for providing me with the amenities to complete my research work. I am grateful to Dr. R V Jasra, Scientist G, for his timely help. I am thankful to Dr. Mrs. K. H. Mody for extending help in conducting some laboratory experiments.

I consider myself profoundly privileged to convey my immense indebtedness to Dr. Anjani Bhatt for her help, advice and scintillating support to fulfill my scientific pursuits. I would also like to express my gargantuan gratitude to all the members of the my discipline as well as administration, library, security, glass blowing and IT cell for their colossal cooperation and appreciation of my exigencies and problems. Undeniably, I am enormously obligated to Department of Biotechnology, New Delhi for its fiscal support.

I am unable to verbalize my delight to thank all my exhilarating and exuberant friends, who ebulliently and efficiently assisted me to tackle my scientific and personal pursuit, and were indeed a moral boost to me. I would like to say a special thanks to Mr. K Ganesan and Mrs. Usha Rani for their warm support.

I am thankful to Lambert Academic Publishing (LAP) GmbH & Co., Germany for publishing my doctoral thesis work in the form of book.

I would like to thank all my family members for their picture perfect encouragement, motivation and indefatigable assistance, which helped me to bring this investigation to a fruitful fruition. I am indeed delighted to fulfill their dreams. Finally, I am grateful to the Almighty who has constantly been with me in my vicissitudes and has brought my doctoral program to a successful completion.

K. Suresh Kumar

Contents

References

Preface

1. Seaweeds

Seaweeds are marine algae – saltwater dwelling, simple multicellular, macrothallic organisms that fall into the rather outdated general category of "plants". They are the oldest members of the plant kingdom, extending back to many hundreds of millions of years. They have little tissue differentiation, no true vascular tissue, no roots and stems, or leaves, and no flowers. They are generally attached by holdfasts that serve anchorage function. They vary in size from small to huge plants more than 100 feet long. They are one of the most important marine renewable and valuable living resources and could be termed "futuristically promising plants of the oceans", with an immense economical and commercial value (Chapman & Chapman, 1980; Levring et al., 1969).

2. Classification of Seaweeds

Seaweeds are classified into three divisions – Chlorophyta (green algae, 1200 species), Phaeophyta (brown algae, 2200 species) and Rhodophyta (red algae, 6500 species) (Smith et al., 1994; Dhargalkar & Pereira, 2005; www.seaweed.ie) based on their pigments and coloration. Other features used to classify them are; cell wall composition, reproductive characteristics, and the chemical nature of their photosynthetic products (oil and starch). Within each of the three major groups of seaweeds, further classification is based on characteristics such as plant structure, form, and shape. Only about 10% of green algae are marine species, and the rest mostly live in freshwater. The green seaweeds are most commonly found in the shallow intertidal zone. More species of green seaweeds are found in warm tropical oceans than in cold temperate seas. Some common species of green seaweeds are *Ulva* (sea lettuce), and *Enteromorpha* (green string lettuce). Brown seaweeds are found in a variety of physical forms including crusts, filaments, and large elaborate kelps. Like all photosynthetic organisms, brown algae contain the green pigment chlorophyll. They also contain other gold and brown pigments, which mask the green color of chlorophyll. The dominant pigment found in brown algae is called fucoxanthin, and it reflects yellow light. Because of the combination of pigments, the coloration of brown algae ranges from light olive green or golden, to very dark brown. Most brown algae live in the mid intertidal or shallow/ upper subtidal zone, and they are most abundant in the colder oceanic waters of the Northern hemisphere. The majority of seaweed species are part of the Division Rhodophyta, or red algae. They have a unique intercellular structure, which give them a rubbery and springy quality. In addition to chlorophyll, red algae contain the pigments phycocyanin and phycoerythrin, which give this group red coloration. However, the colour of red algae varies, and if the pigment phycoerythrin is destroyed, they appear purple, brown, green, or yellow. The accessory pigments (phycocyanin and phycoerythrin) of red algae allow them to grow in deeper waters (subtidal) than other algae. Some red algae also grow in the intertidal zone. One group of red algae, called the coralline algae, are pink in colour and contain deposits of magnesium and calcium carbonate in their cell walls. These seaweeds are hard like stones, and were once thought to be animals closely related to corals (Chapman & Chapman, 1980). Many different uses for red seaweeds have been discovered. Two substances found in the cell walls of red algae are agar and carrageenan. These are gelling compounds, and are used in food products and scientific research. Carrageenan is an important ingredient in toothpaste and many milk products, such as ice cream and chocolate milk. Agar has many scientific applications in microbiology, biotechnology, and criminology, and is used in the packaging of canned meat. One of the most popular red seaweed food is nori (*Porphyra*), which is used in sushi wraps and other Japanese dishes. Nori is grown in commercial seaweed farms on the east coast of North America and in Asia (http://www.oceanlink.island.net/oinfo/seaweeds/seaweeds.html).

3. Production

Currently there are 47 countries in the world with commercial seaweed activity. China holds first rank in seaweed production, with *Laminaria* species. accounting for most of its production, followed by North Korea, South Korea, Japan, Philippines, Chile, Norway, Indonesia, USA and India. According to FAO, between 1981 and 2000, world production of aquatic plants increased from 3.2 million tons to nearly 10.1 million tons (wet weight), upholding US $ 6 billion world trade in 2000 as compared to US $ 250 million trade in 1990. The contribution of cultured seaweeds is 15 % of total global aquaculture (45,715,559 tons). The seaweeds that are mostly exploited for culture are the brown algae with 4,906,280 tons (71 % of total production) followed by the red algae (1,927,917 tons) and green algae (33,700 tons) (Khan & Satam, 2003). Every year about 7.5 – 8 million tons of wet seaweeds are being produced along the coastal regions world wide (McHugh, 2003, 2004). Since 1984, the production of seaweeds worldwide has grown by 119%. The world seaweeds production of 1,338,597 (MT) has been reported for the year 2005 by Bureau of Agricultural Statistics. Seaweeds find their applications as fertilizers and soil conditioners, animal feed, fish feed, biomass for fuel and also in cosmetics, pharmaceuticals, integrated aquaculture and wastewater treatment-phycoremediation (for reduction of nitrogen- and phosphorus-containing compounds and removal of toxic metal).

4. Trends in seaweed research

Since the 1940s, when the potential of agar production from seaweeds was recognized, taxonomy, physiology and biochemistry had been the main research focus. Physiological aspects related to the production of hydrocolloids and pigments, and mass cultivation of seaweeds had been of particular interest. These areas of research laid the foundation of our understanding of seaweed biology. Despite the development and progress of functional genomics in terrestrial plants, seaweeds had received little attention worldwide and were not included in efforts to elucidate gene functions. It was not until the 1990s that studies on the molecular genetics of seaweeds were initiated. These studies were pioneered with the development of genetic transformation techniques on seaweeds and the characterization of genes involved in carbohydrate synthesis. The adaptation of molecular techniques enabled a more reliable and systematic way of inferring evolutionary relationships among different strains or species. Genetic research in seaweeds entered a new phase following the first use of the expressed sequence tag (EST) approach to study seaweed genomics – a relatively inexpensive and quick approach for novel gene discoveries. However, the number of reported seaweed ESTs to date (of which >80% are from the red seaweeds *Porphyra* and *Gracilaria*) accounts for only 0.11% of all publicly available ESTs (Cheong-Xin et al., 2006).

5. Phycocolloids

There are about 9900 species of seaweeds are known and of these only 221 species are used. 101 species (i.e. 68.33 lakh tons) are utilized for phycocolloid production (agar, algin and carrageenan), whereas 145 species are used as food (Zemke–White & Ohno, 1999). The phycocolloid content varies between different species (Black et al., 1951), seasonally (Bird & Hinson, 1992), and among species at different locations (Ohno, Quang, & Hirase 1996; Rebello et al., 1997; Freile-Pelegrin et al., 1996).

6. Agar

Agar is composed of two similar fractions, agarose and agaropectin, in which the basic unit is galactose, linked alternately with α–1,3–(D-galactose) and β–1,4–(α-L-galactose) (Fig. 1). In 1945, Dr. Tseng defined agar as "the dried amorphous, gelatin-like, non-nitrogenous extract from *Gelidium* and other agarophytes, being the sulfuric acid ester of a linear galactan insoluble in cold, but soluble in hot water, and an one per cent neutral solution of which sets at 35°C to 50°C to a firm gel, melting at 80°C to 100°C."

7. Agarophytes

Nearly 33 agarophytic species have been reported till date (Zemke-White, & Ohna, 1999). These include *Gelidium pacificum, Gracilaria* species, *Pterocladia capillace, Ahnpheltia plicata, Acanthopheltis japonica, Ceramiun hypnaeordes* and *Ceranium boydenii*.

8. Applications of agar

About 90 % of the agar produced is used in frozen foods, pastry fillings, syrups, bakery, icings, dry mixes, meringues, frozen desserts, instant pudding, cooked pudding, chiffons, pie and pastry filling, dessert gels; fabricated foods, salad dressings, meat and flavour sauces, food applications, textiles, pharmaceuticals, brewing and cosmetics (McLean et al., 2001). Agar functions as stabilizer, thickener, pre-packing material (for cakes and buns, to reduce the quantity of water), smoothener, non-sticking icing material and in canned products like "scatola" meat (beef blocks in gelatine), as well as in confectionery (jellies, marshmallows and candies as a thickening and gelling agent). In the baked goods industry, the ability of agar gels to withstand high temperatures means agar can be used as a stabilizer and thickener in pie fillings, icings and meringues. Cakes, buns, etc., are often pre-packed in various kinds of modern wrapping materials which often stick to them, especially in hot weather; by reducing the quantity of water and adding some agar, a more stable, smoother, non-stick icing is obtained. In the pharmaceutical industry agar has been used for many years as a smooth laxative. The remaining 10 % is used for bacteriological and other biotechnological uses (Levring et al., 1969).

9. Alginate

Alginic acid is a polyuronide made up of a sequence of two hexuronic acid residues: β–D–mannuronic acid unit and α–L–guluronic acid (Fig. 2).

10. Alginophytes

Most of the large brown seaweeds are potential sources of alginate. There are 41 species of alginophytic seaweeds reported so far. The main commercial sources are species of *Ascophyllum, Durvillaea, Ecklonia, Laminaria, Lessonia, Macrocystis, Sargassum* and *Turbinaria*. Of these the most important are *Macrocystis pyrifera, Laminaria japonica Laminaria hyperborea, Laminaria digitata, Ascophyllum nodosum, Durvillaea lessonia, Durvillaea antarctica, Durvillaea potatorum, Ecklonia cava* and *Eisenia bicyclis* (Zemke–White & Ohno, 1999). *Macrocystis* gives a medium–viscosity alginate, or a high viscosity with a careful extraction procedure (lower temperature for the extraction). *Sargassum* usually gives a low viscosity product. *Laminaria digitata* gives a soft to medium strength gel, while *Laminaria hyperborea* and *Durvillaea* give strong gels. These are some of the reasons why alginate producers like to have a variety of seaweed sources, to produce the desired alginate to the needs of particular applications.

11. Applications of alginate

As a thickener alginate is used in various ways: in ices and ice creams, in products for pastries, in toothpastes, cosmetics, textile printing, paper printing and in water flocculation. On the other hand it is also used as a gelling agent and in entrapment and immobilization of enzymes and microorganisms and also as absorbent products. Alginates improve the texture, body and sheen of yoghurt. It is also used as a suspending agent in chocolate milk and cocoa. Alginates have some applications that are not related to either their viscosity or gel properties. Besides it is also used in the beer and wine industry, in edible dessert jellies, to preserve frozen fish, in meat juices, in pharmaceutical and medical uses, in paper for surface sizing and coating welding rods, as binders for fish feed, as release agents originally for plaster moulds and later in the forming of fibre glass plastics. The Indian alginate industry is based on *Sargassum* species collected from the coasts of Gujarat, Kerala and Tamilnadu. *Sargassum* species obtained from the western coast of India gives a low viscosity alginate, unsuitable for the main Indian market of textile printing. *Turbinaria* is used only when supplies of *Sargassum* are unavailable (Kaliaperumal & Kalimuthu, 1997). The Philippines has large resources of *Sargassum* but this is exported mainly to Japan for use in animal feeds and fertilizers (Chapman & Chapman, 1980; Levring et al., 1969).

12. Carrageenan

Carrageenans are sulfated polymers made up of galactose units (Fig. 3). Several fractions have been determined, but a common backbone can be defined. Carrageenan consists of a main chain of D-galactose residues linked alternately with $\alpha - (1 \rightarrow 3)$ and $\beta - (1 \rightarrow 4)$. The differences between the fractions are due to the number and the position of the sulfate groups and to the possible presence of a 3, 6 anhydro-bridges on the galactose linked through the 1 – and 4 – positions. Different carrageenans are named by Greek letter prefixes: μ, ι, λ, κ, θ, nu and xi-carrageenans (Mueller & Rees, 1968). Three basic types of carrageenans are available: κ–carrageenan from *Chondrus crispus*, *Eucheuma* and *Gigartina* species, ι-carrageenan from *Eucheuma* species and λ- carrageenan from *Chondrus crispus* and *Gigartina* species, which differ in the number and location of sulfate ester substitution. The original classification of carrageenan is determined by the fractionation of the polysaccharide with KCl. The fraction soluble in KCl is called λ-carrageenan and the one insoluble is called κ–carrageenan (Yaphe & Baxter, 1955; Jhonston & McCandless, 1973).

13. Carrageenophytes

The original source of carrageenan is the red seaweed *Chondrus crispus* (common name: Irish Moss), collected from natural resources in France, Ireland, Portugal, Spain and the east coast provinces of Canada. As the carrageenan industry expanded, the demand for raw material had strained the supply from natural resources, although by this time (early 1970s) *Chondrus* was being supplemented by species of *Iridaea* from Chile and *Gigartina* from Spain. The introduction of cultivation of species of *Eucheuma* in the Philippines during the 1970s provided the carrageenan industry with a much enhanced supply of raw material (Doty, 1978).

There are 27 carrageenophytic seaweeds has been reported so far. The most important harvested carrageenophytes are *Sarcothalia crispata*, *Mazzaella laminarioides*, *Gigartina skottsbergii*, *Chondracanthus chamisso* (Norambuena, 1996). Further, carrageenan is also extracted from *Kappaphycus alvarezii* (mainly kappa) and *Eucheuma denticulatum* (mainly iota) (Trono, 1993). *Gigartina skottsbergii* (mainly kappa, some lambda), *Sarcothalia crispate* (mixture of kappa and lambda) and *Mazzaella laminaroides* are currently the most valuable species, all collected from natural resources in Chile. Small quantities of *Gigartina canaliculata* are harvested

5

in Mexico. *Hypnea musciformis* has been used in Brazil (Bravin and Yoneshigue-Valentin, 2002). In India, *Hypnea musciformis* and *H. valentiae* are commercially exploited carrageenophytes (Ramalingam et al., 2003). A few of the carrageenophytic seaweeds are shown in fig. 4.

The estimated amount of carrageenophytes being produced yearly has crossed 81,858 t (d. wt) with carrageenan production of about 28,650 t. This resource costs 25 kg $^{-1}$ US $ and has an approximate annual value of US $ 2.6 billion (Chopin et al., 1995; Ohno et al., 1996).

14. Applications of carrageenan

Carrageenan is mainly employed in food industry especially in the manufacture of sausages, corned beef, meat balls, ham, preparations of poultry and fish, chocolate, dessert gel, ice cream, juice concentrates, marmalade, sardine sauce fruit. It is also used in gelation, fat stabilization, thickening, suspending gelation, suspension, bodying, pulping effects, emulsion stabilization, binder, emollient. Its non–food applications include its use in manufacturing beer, air fresheners, textiles, toothpastes, hair shampoo, sanitary napkins, tissue, culture media, fungicide, etc. (Moirano, 1977).

15. Taxonomy of *Kappaphycus alvarezii* (www.algaebase.org)

Kingdom: Plantae
    Subkingdom: Biliphyta
        Phylum: Rhodophyta
            Subphylum: Rhodophytina
                Class: Florideophyceae
                    Subclass: Rhodymeniophycidae
                        Order: Gigartinales
                            Family: Areschougiaceae
                                Tribe: Eucheumatoideae
                                    Genus: *Kappaphycus*
                          Species: *alvarezii* (Doty) Doty ex P.C. Silva
                Botanical Name: *Kappaphycus alvarezii* (Doty) Doty ex P.C. Silva

Three commercial species of *Eucheuma* are being used for their carrageenans. The annual production of carrageenans of each species is: *Eucheuma cottonii*, 30,000 tons producing kappa carrageenan; *Eucheuma spinosum*, over 6,000 tons producing iota carrageenan; and *Eucheuma gelatinae*, about 100 tons producing a mixture of gamma, beta and kappa carrageenans. Over 95% of the annual commercial *Eucheuma* crop is from farms in the tropical far western Pacific. In addition to foreign exchange earnings for those countries exporting the seaweed, the labor-intensive farming of *Eucheuma* is of great socio-economic value to the often nearly indigent shore dwelling families who grow it.

6

The two species originally cultivated in the Philippines were *Eucheuma cottonii* and *Eucheuma spinosum*, and often referred to as "cottonii" and "spinosum" respectively. However, phycologists have since renamed both species: *Eucheuma cottonii* as *Kappaphycus alvarezii*, and *Eucheuma spinosum* as *Eucheuma denticulatum*. Unfortunately all the names are still in use and so an awareness of them is necessary while reading about carrageenophytes (FAO, 2003). *Kappaphycus* species are among the largest tropical red seaweed, with a high growth rate. *Kappaphycus alvarezii* is amongst the most widely studied edible seaweed (Trono, 1993). Its taxonomy has been studied by Doty (1988).

16. Life History

The reproduction of *Eucheuma* species from a classical point of view is reported by Kylin (1956), Gabrielson (1983) and Gabrielson & Kraft (1984). The nature of its triphasic life history results in its relation to the family Solieriaceae of the order Gigartinales. A diploid vegetative phase (asexual) produces haploid non-motile spores called tetra–spores. The tetraspores produce haploid gametophytes (sexual) which in turn produce diploid carposporophytes that are parasitic in the female thalli. The carposporophytes release diploid carpospores which initiate the diploid tetrasporic stage again. Superfically spinosum, cottonii and gelatinae all appear to be triphasic though the strains being farmed may not be. There is no detailed study of the life history of any member of this genus. Male thalli are as yet unknown for the commercial forms. Santos & Doty (1978) found both cystocarpic and tetrasporic thalli in quantity in only 6 of the 15 species in which they were sought. They did not find them common in the commercial species and found no male thalli. *Eucheuma alvarezii* (=*Kappaphycus alvarezii*) carpospores produce morula-like tetra-sporophytic embryoes that attach by rhizoids and differentiate into an erect pyriform followed by cylindrical form as the typical cluster of apical cells develops opposite the substratum. The younger embryos have a thick gel coat and a large number of very fine hairs, each ten or twenty times longer than the diameter of the embryo. Perhaps they may represent the micro-form in which some species of the genus persist in "off" seasons (Santos & Doty, 1978).

17. Morphology and Anatomy

Kappaphycus alvarezii plants are tough, fleshy, firm; and their length can extend even up to 2 m. The thalli are coarse, with axes and branches 1 – 2 cm diameter; heavy, with major axes relatively straight, lacking secondary branches near apices. It is frequently and irregularly branched, most branches primary, secondary branches intercalated between primary branches or mostly lacking. The plants contain a few small branches in shallow areas and in deeper waters they are large, intricately tangled as fleshy mats.

The thalli of commercial *Kappaphycus alvarezii* are often grow up to a kilogram in mass. Basically, the thallus is a multiaxial filamentous red algal genus which becomes strongly pseudoparenchymatous. The species vary greatly in form as a result of the environments in which they grow. This is particularly true of those in the section of the genus to which the *cottonii* forms belong. Commercial *spinosum* and *cottonii* are composed of cylindrical branches that are rigid. Mature *gelatinae* branches are apically flat, somewhat flexible, have marginal teeth and arise from a cushion of strongly compressed branches. Gabrielson (1983) and Gabrielson & Kraft (1984) provided much of the structural details which Kylin (1956) did not include. None of these authors treated the sections of the genus, *Anaxiferae, Cottoniformia* or *Gelatiformia*, from which the commercial *cottonii* and *gelatinae* crops arise. Weber-van Bosse (1928) and Doty & Norris (1985) provided some of the structural details of the section *Cottoniformia* but they were still largely undefined anatomically. The different color forms of *Kappaphycus alvarezii* such as brown, green and pale yellow forms show typical variation in their morphological features. The brown color forms are dark brown in color, branches are thick and robust, profusely and irregularly branched, the branches are tapering towards the tips. The basal part of the branch is very thick; plant normally measures up to 20 – 25 cm. Green color forms are dark green in color, branches are thick and profusely branched with fragile tips; plants are smaller than the brown form and measure up to 20 cm. Pale yellow form are light yellowish in color, smaller than the other two color forms, less branched. Branches very fragile and break even at moderate water current; the plant measures up to 14 – 16 cm. (Suresh Kumar *et al.,* 2007 b)

18. Habitat

The plants occur in reef flat and reef edge, 1 to 17 m deep. Loosely attached to broken coral, or unattached fragments floating in shallow and deep waters can form large, moving mats of unattached thalli (Trono, 1993).

19. Geographical distribution

The two major commercial forms, *spinosum* and *cottonii*, are native to the Old World Tropics (Weber-van Bosse 1928; Laite & Ricohermoso 1981) westward to East Africa (Anderson 1953; Mshigeni 1982, 1984). A very small amount of the third, *gelatinae*, comes from the Philippines and China (Hainan Island and Taiwan). These three species of the genus are almost entirely restricted to the same brightly lit waters in which coral reefs form. *Eucheuma* is now being commercially produced in the far western Pacific. *Cottonii* has been introduced eastward through Micronesia into Kiribati and Tonga to the Society and Hawaiian Islands. *Kappaphycus alvarezii* has been introduced into Fiji and Hawaiian Islands, and also in Indonesia and China. *K. cottonii* is harvested along the coasts of Indonesia, Kenya, Tanzania, China, Japan, Vietnam, Malaysia, Guam and Fiji. On the other hand *K. striatum* is reported from the waters of Fiji and Hawaiian Islands, Kenya, Tanzania, Madagascar, Japan, China, Singapore and Indonesia (Guiry & Dhonncha, 2005).

20. Study of *Kappaphycus avlaverzii*: Status today

Until recently in Hawaii, *Kappaphycus alvarezii* is found to reproduce sexually. Since its introduction to other localities, observations and studies have reported the propogation by vegetative fragmentation. At the tip of each branch is a cluster of apical cells potentially high in regenerative capabilities that are able to regenerate a new thallus after breaking off. A broken tip can grow into full-sized thalli in a short period of time. This species has been highly successful at Kaneohe Bay, dominating the sandy spur and grooves on the reef flat. It inhabits barren sandy grooves where it does not appear to compete with native algal species. The red alga's dispersal is thought to be constrained by size and weight, as it appears to become trapped in depressions and channels. The species is also constrained by high herbivory. *K. alvarezii* has managed to spread to neighboring reefs with supportive physical factors and little grazing, where it is dominating the changing marine ecology (www.hawaii.edu/reefalgae/invasive_algae/rhodo/kappaphycus_alvarezii.htm). It has been introduced throughout the warm tropics for commercial cultivation (c.f. Hurtado et al., 2001). It is a major producer of kappa–carrageenan, besides being a source of human food (Trono, 1993).

Apart from being used in various commercial products such as toothpaste, ice-cream, pet food and soft capsules, Central Salt and Marine Chemicals Research Institute (CSMCRI), Bhavnagar has been instrumental in obtaining multiple products from *Kappaphycus alvarezii*. This process includes the following steps

- Obtaining liquid seaweed fertilizer by crushing the fresh algae, Moreover, the granules obtained after extraction of LSF can be utilized for preparation of both refined and semi refined carrageenan (US Patent No. 6893479)
- Alternatively, the fresh algae when dried yields a low sodium salt – Saloni K (US Patent No. 7,208,189).
- Health drink prepared from the seaweed sap (Indian Patent No. 134 NF/2006).
- Dried seaweed powder used as human food (Indian Patent No. 143 NF /2005).

On 11 May 2006, Dr. A.P.J. Abdul Kalam the former President of India addressed the nation on the occasion of Technology day, wherein he mentioned about this integrated technology. He also mentioned about employment genereated for over 2000 families with an average income of over Rs. 6000 per month, as 40,000 rafts were placed in the sea at Mandapam, Tamilnadu for cultivation. Its estimated number will go up to 1.5 lakhs by the end of 2007.

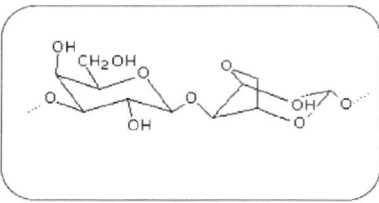

Fig. 1. Structure of agar

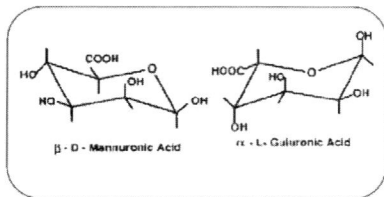

Fig. 2. Structure of alginate

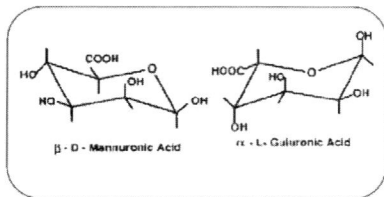

Fig. 3. Types of carrageenans (structure)

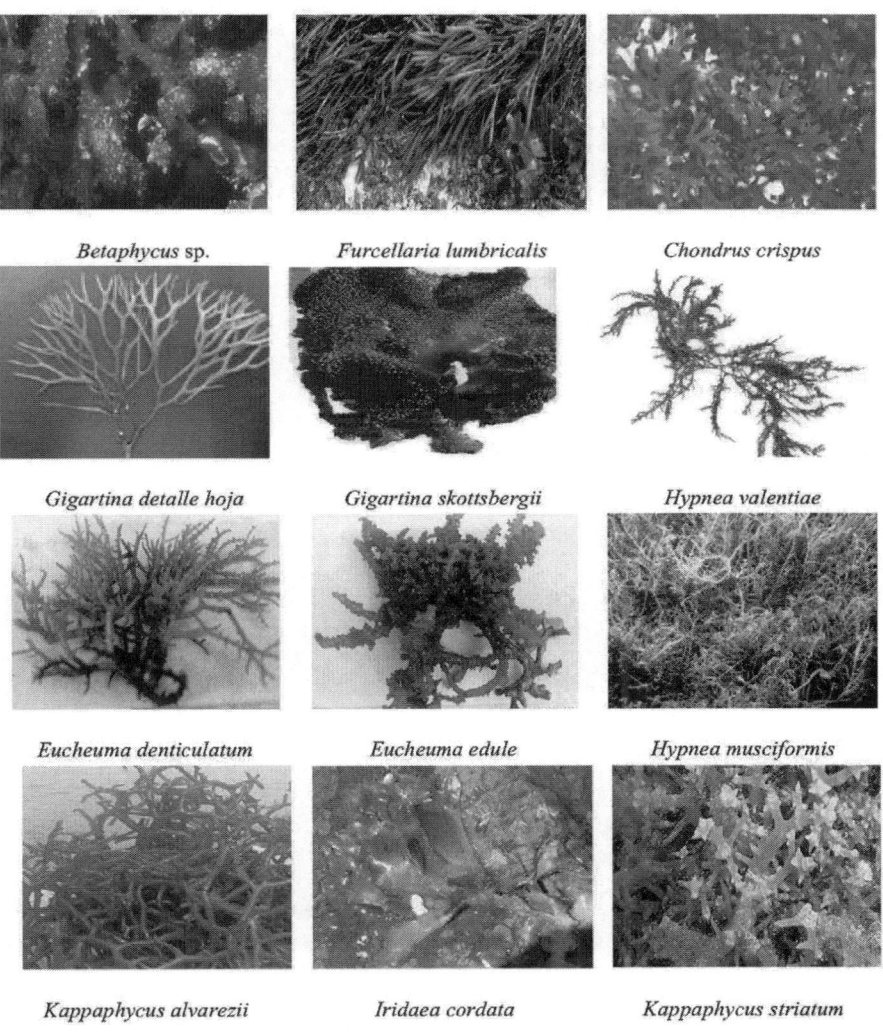

Fig. 4. A few carrageenophytes

11

References

Amutha, S., Bhat, K.K., Ravishankar, G.A., & Subba Rao, P.V. (2005). A Process for the preparation of Eucheuma powder (Kappaphycus alvarezii) an edible seaweed for use in food formulations.Indian Patent No 143 NF /2005.

Anderson, G.W. (1953). A note on the seaweed resources of Zanzibar protectorate. *Proceeding of International Seaweed Symposium, 1,*102–103.

Bird, K.T., & Hinson, T.K. (1992). Seasonal variations in agar yeilds and quality from North Carolina agarophytes. *Botanica marina, 35,* 291–295.

Black, W.A.P., Dewar, E.T., & Woodward, F.N. (1951). Manufacture of algal chemicals. II. Laboratory-scale isolation of mannitol from brown marine algae. *Journal of Applied Chemistry, 1,* 414–424.

Bravin, I.C., & Yoneshigue-Valentin, Y. (2002). The influence of environemtal factors on *in vitro* growth of *Hypnea musciformis* (Wulfen) Lamouroux (Rhodophyta). *Reviews in Brazilian Botany, 25,* 469-474.

Bureau of Agricultural Statistics. (2005). News article by Benjamin V Bucor Jr., Economic Indicators, including data from Bureau of Agricultural Statistics of November 29, 2006, Manila, Philippines, http://www.bworld.com.ph/Research/economicindicators as on 02-06-2007.

Chapman, V.J., & Chapman, D.J. (1980). Seaweeds and their uses. Chapman & Hall, London.

Cheong-Xin C., Chai-Ling H., & Siew-Moi P. (2006). Trends in seaweed research. *Trends in Plant Science, 11 (4),* 165–166.

Chopin, T., Gallant, T., & Davison, I. (1995). Phosphorus and nitrogen nutrition in *Chondrus crispus* (Rhodophyta): Effects on total phosphorus and nitrogen content, carrageenan production, and photosynthetic pigments and metabolism. *Journal of Phycology, 31,* 283–293.

Dhargalkar, V.K., & Pereira, N. (2005). Seaweed: promising plant of the millennium, *Science and Culture*, March-April, *4,* 60-66.

Doty, M. S. (1988). Prodromus ad systematica Eucheumatoideorum: a tribe of commercial seaweedsrelated to *Eucheuma* (Solieriaceae, Gigartinales). pp. 159–207. In Abbott I. A. (ed.). Taxonomy of economic seaweeds with special reference to Pacific and Caribbean species. Volume 11. CaliforniaSea Grant College Program Rep. No. T-CSGCP-018.

Doty, M.S., & Norris, J.N. (1985). *Eucheuma* species (Solieriaceae, Rhodophyta) that are major sources of carrageenan. *In I.A. Abbott & J.N. Norris (eds), Taxonomy of Economic Seaweeds. California Sea Grant College Program Report No.T-CSGCP-011,* 47–61.

Eswaran, K., Ghosh, P.K., Siddhanta., Arup, K., Patolia, J.S., Periyasamy., Chellaiah., Mehta, A.S., Mody, K.H., Ramavat, B.K., Kamalesh., Rajyaguru., Rameshchandra., Reddy, K.C.R., Pandya, J.B., & Tewar, A. (2005).Integrated method for production of carrageenan and liquid fertilizer from fresh seaweeds USP 6893479.

FAO. (2001). The state of world fisheries and aquaculture 2000, http://fao.org.

FAO. (2003). A guide to the seaweed industry, Edited by Dennis J. McHugh, FAO Fisheries Technical Papers Series - T441, ftp://ftp.fao.org/docrep/fao/006/y4765e/y4765e00.pdf, ISBN: 9251049580, 51–60.

Freile-Pelegrin, Y., Robledo, D.R., & Garcia-Reina, G. (1995). Seasonal agar yield and quality in *Gelidium canariensis* (Grunow) *Seoane Camba* (Gelidiales, Rhodophyta) from Gran Canaria, Spain. *Journal of Applied Phycology, 7,* 141–144.

Gabrielson, P. W., & G. T. Kraft. (1984). The marine algae of Lord Howe Island (N.S.W.): the family Solieriaceae (Gigartinales, Rhodophyta). *Brunonia 7,* 217–251.

Gabrielson, P.W. (1983). Vegetative and reproductive morphology of *Eucheuma isiforme*(Solieriaceae, Gigartinales, Rhodophyta). *Journal of Phycology, 19,* 45–52.

Ghosh, P.K., Mody, K.H., Reddy, M.P., Patolia, J.S., Eswaran, K., Rajul Shah., Bhargav Barot., Gandhi, M.R., Mehta, A.S., Bhatt, A.M., & Reddy, A.V.R (2007). Low sodium salt of botanic origin, US patent No. 7,208,189 dated 24 April 2007. Also published as WO200509768.

Ghosh, P.K., Rajyaguru, M.R., Patolia, J.S., Eswaran, K., Subba Rao, P.V., Shah, M.T., Zodape, S.T., Joshi, S.V., Reddy, A.V.R., Devmurari, C.V., Bandyopadhyay, S., & Sahoo, G.C. (2006). Refreshing and tasty drink and a process for the preparation thereof. Indian Patent No .134 NF/2006.

Guiry, M.D., & Dhonncha E.N. (2005). *AlgaeBase version 2.1.* Worldwide electronic publication, National University of Ireland. Galway: http://www.algaebase.org (15 February 2005).

http://www.hawaii.edu/reefalgae/invasive_algae/rhodo/kappaphycus_alvarezii.htm as on 4th September 2007.

http://www.oceanlink.island.net/oinfo/seaweeds/seaweeds.html as on 4th September 2007.

Johnston, K.H., & McCandless, E.L. (1973). Enzymic hydrolysis of the potassium chloride soluble fraction of carrageenan: Properties of "λ-carrageenase" from *Pseudomonas carrageenovora*. *Canadian Journal of Microbiology, 19,* 779–788.

Kaliaperumal, N., & Kalimuthu, S. (1997). Seaweed potential and its exploitation in India. *Seaweed Research and Utilization, 19,* 33-40.

Khan, S.I., & Satam, S.B. (2003). Seaweed Mariculture: Scope and Potential in India, *Aquaculture Asia, 8(4),* 426-29.

Kylin, H. (1956). Die Gattungen der Rhodophyceen. Lund, K. Gleerups, 673 .

Laite, P., & Riconermoso, M. (1981). Revolutionary impact of *Eucheuma* cultivation in the South China Sea on the carrageenan industry. *Proceeding of .International Seaweed Symposium, 10,* 596–600.

Levring, T., Hoppe, H.A., &. Schmid, O.J. (1969). Marine algae. A survey of research and utilization. Cram de Grutjter. *1,* 421.

McHugh, D.J. (2002). Prospects for seaweed production in developing countries. FAO Fisheries Circular. No. 968. Rome, FAO. pp. 28.

McHugh, D.J. (2003). A guide to the seaweed industry. FAO Fisheries Technical Paper 441, pp. 105.

McLean, E., Craig, S., & Schwartz, M. (2001). Macrophyte biotechnology: a brief review of production methods and industrial application of seaweed. *Aquaculture, 27,* 341–358.

Moirano, A. (1977) Sulfated seaweed polysaccharides. In: Food colloids, Edited by Graham H.D. Westport, Connecticut, AVI Publishing Co., Inc., 347–81.

Mshigeni, K.E. (1984). The red algal genus *Eucheuma* (Gigartinales Solieraceae) in East Africa: and an underexploited resource. *Hydrobiologia, 116/117,* 347–50.

Mshigeni, K.E. (1982). Seaweed resources in Tanzania: a survey of potential sources for industrial phycocolloids and for other uses. In: Marine algae in pharmaceutical science, 2, Berlin, edited by. Hoppe, H.A., &. Levring, T. Walter de Gruyter and Co., pp. 174.

Mueller, G.P., & Rees, D.A. (1968). Current structural views of red seaweed polysaccharides. In Drugs from the sea, edited by Freundenthal H.D., Washington, D.C., *Marine Technology Society*, pp.241–55.

Norambuena, R. (1996). Recent trends of seaweed production in Chile. *Hydrobiologia 326/327,* 371–379.

Ohno, M., Quang, N.H., & Hirase, S. (1996). Cultivation and carrageenan yield and quality of *Kappaphycus alvarezii* in the waters of Vietnam. *Journal of Applied Phycology, 8,* 431–437.

Ramalingam, J.R., Kaliaperumal, N., & Kalimuthu, S. (2003). Commercial scale production of carrageenan from red algae. *Seaweed Research and Utilization, 25,* 37–46.

Rebello, J., Ohno, M., Ukeda, H., & Sawamura, M. (1997). Agar quality of commercial agarophytes from different geographical origins: 1. Physical and rheological properties. *Journal of Applied Phycology, 8,* 517–521.

Santos, G.A., & M.S. Doty. (1978). *Gracilaria* for the manufacture of agar. *Fisheries Research Journal Philippines, 3,* 29–34.

Smith, G. M. (1944). *Marine Algae of the Monterey Peninsula*, Stanford Univ., California, 2nd Ed.

Subba Rao, P.V., & Mantri V.A. (2006). Indian seaweed resources and sustainable utilization: Scenario at the dawn of a new century, *Current Science, 91(2),* 164–174.

Suresh Kumar, K., Ganesan, K, & Subba Rao P. V. (2007). Heavy metal chelation by non-living biomass of three color forms of *Kappaphycus alvarezii* (Doty) Doty. DOI 10.1007/s10811-007-9181-8.

Trono, G.C. (1993). *Eucheuma* and *Kappaphycus*: Taxonomy and cultivation. In Seaweed cultivation and marine ranching (eds.) Ohno, M., & Critchley, Japan International Corporation Agency (JICA), Yokosuka, Japan. pp. 75–88.

Weber-van Bosse, A. (1928). Liste des algues du Siboga. IV. Part 3: Gigartinales et Rhodymeniales. In *M. Weber (ed.), Siboga Expedie, Monog. 59d*, pp. 393–533.

www.seaweed.ie as on 4th September 2007.

Yaphe, W., & Baxter, B. (1955). The Enzymic Hydrolysis of Carrageenin. *Applied and Environment Microbiology, 3,* 380–383.

Zemke-White, W.L., & Ohno, M. (1999).World seaweed utilisation: An end-of-century summary. *Journal of Applied Phycology, 11,* 369–376.

Chapter I
Cultivation of Kappaphycus alvarezii

1. Introduction

Kappaphycus alvarezii is economically important tropical red seaweed with a high demand worldwide for its cell wall polysaccharide, carrageenan, (Bixler1996). Commercial cultivation of *K. alvarezii* originated in Philippines during the latter half of the 1960s (Doty 1973; Parker 1974; Doty & Alvarez 1975). Current sources of cultivated eucheumatoids seem incapable of meeting its growing market demand of carrageenan (Ask & Azanza 2002), and in order to meet the demand the cultivation of *Kappaphycus* species has expanded to other regions of Philippines and also to other countries like Japan and Indonesia (Adnan & Porse 1987; Luxton 1993), Tanzania (Lirasan & Twide 1993), Kiripati (Luxton & Luxton 1999), Fiji (Luxton, Robertson & Kindley 1987), Hawaii (Glenn & Doty 1990) and South Africa (Braud & Perez 1978; Lirasan & Twide 1993). Experimental cultivation of this seaweed has been accomplished successfully in Madagascar (Mollion & Braud 1993), Vietnam (Ohno, Nang & Hirase 1996) and China (Qian, Wu, Wu & Xie 1996). Warm, nutrient-enriched seawater, high light levels and high degree of water motion facilitate successful cultivation of Kappaphy- cus species (Glenn & Doty 1990). Some preliminary experiments in tide pools on field cultivation of this seaweed (5Kappaphycus striatum) were carried out at Okha, Northwest coast of India (Mairh, Zodape, Tewari & Rajyaguru 1995). The cultivation of this sea- weed was initially attempted at Mandapam on the Southeast coast of India (Eswaran, Ghosh & Mairh 2002). In the present study, an attempt was made to cultivate this seaweed in the open sea at three local- ities, viz., Mithapur, Okha and Beyt Dwaraka on the Northwest coast of the Indian Peninsula, and this study reports the results obtained, providing an over- all picture to further take up large-scale cultivation for the benefit of the coastal rural folk.

2. Materials and methods

Kappaphycus alvarezii (Doty) Doty seed material was drawn from the cultivation farm of Diu (and Diu farm material originally originated from the cultivation farm of Mandapam, Southeast coast of Indian Peninsula). Initially, 10 kg material was collected from the Diu farm and domesticated at Okha during August and September 2004 to obtain sufficient seed material for further cultivation. The cultivation at Mithapur ($22°17.630'N$ and $69°58.22'E$), Okha ($22°28.656'N$ and $69°04.015'E$) and Beyt Dwaraka ($22°26.38'N$ and $69°03.35'E$) (Fig. 1) was carried out from September 2004 to March 2005 to evaluate the growth in terms of Biomass (fresh) and daily growth rate (DGR).

Figure 1 Map showing the location of experimental sites on the Northwest coast, India.

The cultivation was carried out using the raft method (Trono & Ohno 1989). The size of the raft (2.5 m × 2.5 m) was chosen to suit the experimental site. The raft contained 16 monolines of 3 mm polypropylene rope of 2.5 m length at 15 cm intervals. Each line consisted of 16 seedlings of 100 g (fresh weight) each at 15-cm intervals and thus each line contained (100 g × 16) = 1.6 kg (fresh weight) seed material. The seedling was tied to a monoline using the tie–tie method (Doty & Alvarez 1975). Thus, each raft contained (1.6 × 16) = 25.6 kg (fresh weight) seed material and five such rafts were used at each experimental site and they were anchored in the sea so that they were not normally exposed in the low tides. Harvest of the five plants at random was made from each raft at every 15–day interval up to 45 days i.e. after15, 30 and 45 days of planting. The biomass of the harvested plants was measured and the DGR% was calculated using the formula adopted by Dawes, Lluisma and Trono (1994).

$$DGR \% = \ln (W_f / W_0) / t \times 100$$

Where W_f is the final fresh weight (g) at t day, W_0 is the initial fresh weight (g), t is the number of culture days. The seed material for each plantation was obtained from preceding harvested materials.

2.1. Environmental factors

Seawater temperature was recorded in the vicinity of experimental sites at fortnightly intervals. The seawater water samples were collected from the cultivation sites at fortnightly intervals and were analysed in duplicate for salinity and nutrients, viz., nitrate and phosphate following the methods described by Strickland and Parsons (1972). The average values for seawater temperature, salinity, nitrate and phosphate for each month were finally obtained.

2.2. Extraction of semi-refined carrageenan

The pre-cleaned dried seaweed (10 g) was rinsed with freshwater at ambient room temperature and treated with 200mL of 8% KOH (cooking solution) maintained at approximately 70 °C. This process enabled modification of the carrageenan and dissolved some of the alkali-soluble sugars (and modified others to become soluble in water).This KOH mixture, containing seaweed at a pH between12 and14, was very corrosive and required to be handled with extreme caution. After the cooking time (3 h) had elapsed, the seaweed was removed from the hot aqueous KOH solution. Later, the seaweed was subjected to a series of wash steps to reduce the pH, by washing residual KOH from the seaweed. The semi-refined carrageenan was obtained after repeated washing the seaweed and finally it was dried and ground (Christopher & Michael 1997).

2.3. Statistical analysis

The biomass (fresh) and growth rates were expressed in terms of mean standard deviation. In addition, Pearson's correlation coefficient (r) was computed for growth rates obtained for a 30-day growth period and environmental parameters. One–way ANOVA was adopted to find out significant differences among the growth rates for 30-day grown plants of three sites.

4. Results

Monthly changes in environmental factors at the experimental sites (Mithapur, Okha and Beyt Dwarka) are shown in Fig. 2a and b. Seawater temperature varied from 19.9 °C in January 2005 to 26.5 °C in October 2004 at Mithapur, from 19.3 °C in December 2004 to 26.9 °C in March 2005 at Okha and from 19.8 °C in January 2005 to 25.4 °C in March 2005 at Beyt Dwarka. The salinity (‰) was found to vary from 30.81 in November 2004 to 36.83 in March 2005 at Mithapur, from 30.96 in December 2004 to 36.12 in March 2005 at Okha and from 30.86 in November 2004 to 36.01 in March 2005 at Beyt Dwarka. (Fig. 2a). The nutrients ranged from 11.90 in March to 19.75 mmol L^{-1} in October, from 11.64 in February to 18.98 mmol L^{-1} in October and from 12.16 in December to 19.54 mmol L^{-1} in October, for nitrate at Mithapur, Okha and Beyt Dwarka, respectively, with respective values of 2.91 in January to 5.00 mmol L^{-1} in October, 2.37 in January to 3.64 mmol L^{-1} in October and 2.62 in January to 3.91 mmol L^{-1} in March for phosphate (Fig. 2b). Generally, seawater at these experimental sites showed high contents of nitrate and phosphate.

17

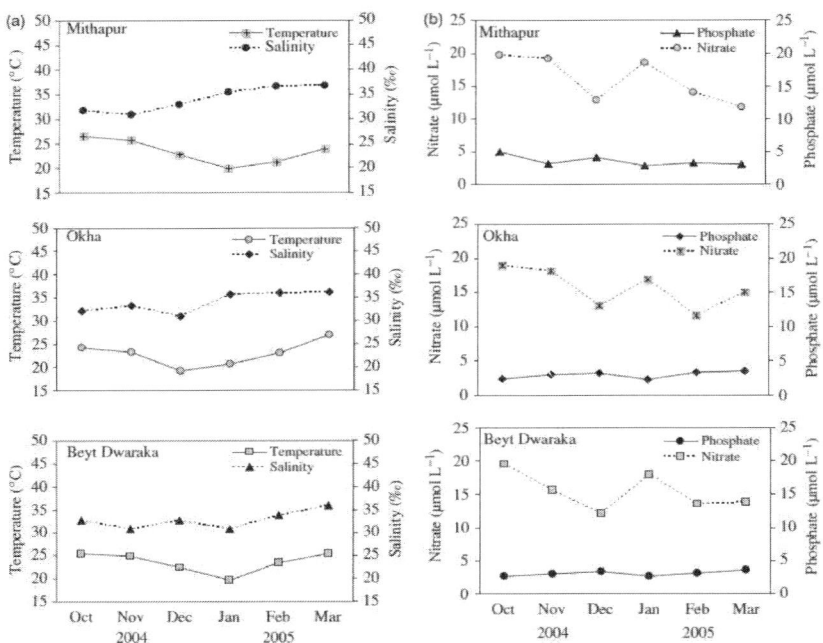

Figure 2 (a) Monthly variation in temperature and salinity at three cultivation sites. (b) Monthly variation in nitrate and phosphate at three cultivation sites.

Seasonal variations were observed in biomass and growth rates at 15, 30 and 45 days (Fig. 3a and b). During the 15-day growth period, the minimum biomass of 252.0 ± 118.35 g in November and the maximum biomass 435.0 ± 23.66 g in October were observed at Mithapur and at Okha it was minimum 189.0 ± 10.20 g in December and maximum 401.0 ± 30.89 g in October and at Beyt Dwarka the same was mini- mum 182.0 ± 19.96 g in December and maximum 407.0 ± 21.35 g in October (Fig. 3a). During this period, the DGR (%) was found to vary from 4.05 ± 0.66 in October to 9.80 ± 0.63 in March at Mithapur, from 4.23 ± 0.36 in October to 9.24 ± 0.53 in March at Okha and from 3.95 ± 0.78 in October to 9.35 ± 0.36 in March at Beyt Dwarka (Fig. 3b). The minimum and maximum growth rates were recorded in October and March, respectively, at all the places.

During the 30-day growth period, the minimum and maximum biomass were recorded in November (326.0 ± 27.82 g) and in March (1096.0 ± 61.43 g), respectively, at Mithapur, whereas the respective values   were   376.0 ± 19.33 g   in   December and 1033.5 ± 46.06 g in October at Okha and with corre- sponding values of 306.0 ± 118.09 g in November and 814.0 ± 76.84 g in

18

October at Beyt Dwarka (Fig. 3a). During the same period the DGR (%) varied from 3.93 ± 0.29 in November to 7.98 ± 0.19 in March at Mithapur, from 7.93 ± 1.78 in December to 13.98 ±

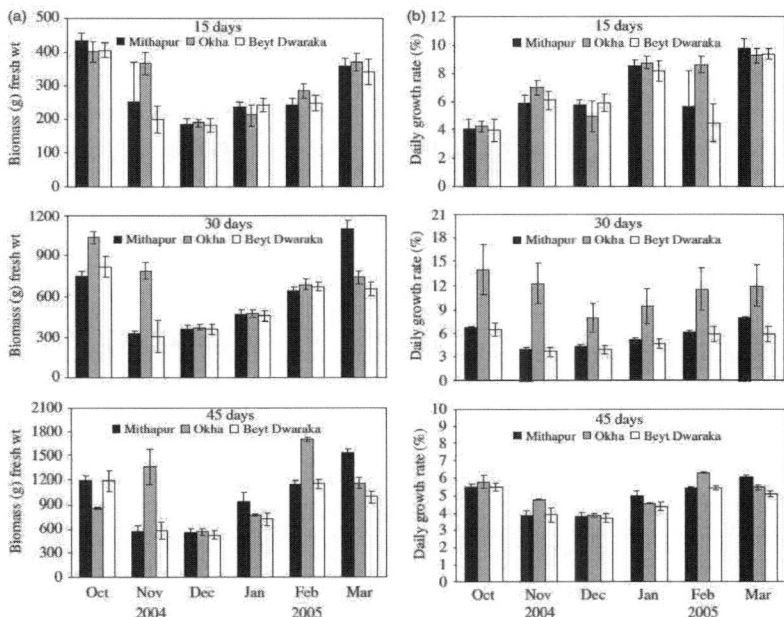

Figure 3 (a) Biomass yields (15, 30 and 45 days) at different cultivation sites. (b) Daily growth rates (15, 30 and 45 days) at different cultivation sites.

3.07 in October at Okha and from 3.64 ± 0.01 in November to 6.49 ± 0.89 in October at Beyt Dwarka (Fig. 3b).

During the 45-day growth period, a minimum biomass of 553.0 ± 53.14 g in December and a maximum biomass 1537.0 ± 43.54 g in March were recorded at Mithapur with respective values of 569.0 ± 37.33 g in December and 1363.5 ± 219.12 g in October at Okha and with corresponding values of 530.5 ± 50.95 g in December and 1193.0 ± 123.76 g in October at Beyt Dwarka (Fig. 3a). During this period, the DGR (%) varied from 3.79 ± 0.23 in December to 6.07 ± 0.06 in March at Mithapur, from 3.86 ± 0.14 in December to 5.78 ± 0.35 in October at Okha and from 3.69 ± 0.23 in December to 5.43 ± 0.12 in February at Beyt Dwarka (Fig. 3b). A high deviation was found in some cases (for minimum biomass of 15 days growth at Mithapur and for minimum biomass of 30 days growth at Beyt Dwarka) because there was loss of branches of the plants taken for weights due to severe wave action coupled with high-velocity winds.

The semi-refined carrageenan yields were found to vary from 44.92 ± 0.33% to 53.74 ± 3.45%, from 52.57 ± 0.73% to 58.36 ± 1.26% and from 42.42 ± 1.89% to 51.70 ± 2.25% at Mithpur, Okha and Beyt Dwaraka respectively (Fig. 4).

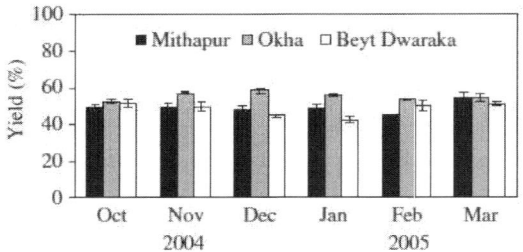

Figure 4 Semi-refined carrageenan yield (%) at different cultivation sites.

5. Discussion

Kappaphycus alvarezii was grown in subtropical waters of Northwest coast - Mithapur, Okha and Beyt Dwarka – of the Indian Peninsula during September to March. The maximum growth rates obtained were 9.35-13.98% and these growth rates were comparable with the growth rates reported elsewhere (Table1).

Ohno, Largo and Ikurnoto (1994) recorded a growth rate of 0.13 – 8.12% for this seaweed in subtropical waters of Shikoku, Japan. Further, at Nha Trang, Vietnam, growth rates of 3.95 – 10.80% (Ohno, Nang, Dinh & Triet 1995) and 3.16 – 10.80% (Ohno et al. 1996) were also recorded for the same seaweed. Paula, Pereira and Ohno (2002) obtained a growth rate of 4.5 – 8.2% in subtropical waters of Brazil. In the tropical waters of Yucatan peninsula, Mexico, the growth rate was different (2.0 – 7.1%) (Muñoz, Freile-Pelegrín & Robledo 2004). Hurtado, Agbayani, Sanares and Castro–Mallare (2001) reported a growth rate of 2.3 – 4.2% for this seaweed cultivated in Pangatan Cays, Caluya, Antique, Philippines, while Glenn and Doty (1990) reported a growth rate of 1.9 – 6.2% in Hawaii and in Dzilam, Yucatan, Mexico, the growth rate recorded was 2.0 – 3.3% for brown, green and red strains of this seaweed. Dawes et al. (1994) reported a growth rate of 5.9 – 8.9% in Philippines. In an earlier preliminary study, the growth rates recorded for this seaweed in Indian subtropical waters were 2.45 – 7.64% at Okha, Northwest coast of India (Mairh et al. 1995), and 0.34 – 5.5% at Mandapam, Southeast coast of India (Eswaran et al. 2002) and these growth rates were found to be less than the ones recorded in the present study (3.64 ± 0.01 – 13.98 ± 3.07%). However, the growth rate at Okha was the highest reported so far for this seaweed and it was significant at Mithapur and Okha and non-significant at

Beyt Dwarka ($p < 0.01$). At Mithapur, the growth rate was significantly correlated with salinity ($r = 0.659$) at $p < 0.05$. At Okha and Beyt Dwarka, it was significantly correlated with nitrate ($r = 0.714$ at Okha and $r = 0.50$ at Beyt Dwarka) and with seawater temperature ($r = 0.829$ at Okha and $r = 0.510$ at Beyt Dwarka) at $p < 0.05$.

Table 1 Daily growth rate of *Kappaphycus alvarezii* reported in tropical and subtropical waters

Location	Growth rate (% day^{-1})	Authors
Philippines	4.8–5.6	Doty and Alvarez (1981)
Hawaii	3.5	Glenn and Doty (1981)
Indonesia	3–4	Adnan and Porse (1987)
Fiji	2.3–5.3	Luxton et al. (1987)
Maldives	2.0–7.0	De Reviers (1989)
Hawaii	1.9–6.2	Glenn and Doty (1990)
Dzilam	2.0–3.3	Glenn and Doty (1990)
China	1.9–4.6	Li, Li and Wu (1990)
Hawaii	1.5–5.6	Glenn and Doty (1992)
Philippines	3.7–7.2	Hurtado-Ponce (1992)
Zanzibar	5–7	Lirasan and Twide (1993)
Philippines	5.9–8.9	Dawes et al. (1994)
Shikoku Japan	0.13–8.12	Ohno et al. (1994)
Philippines	1.1–3.4	Hurtado (1995)
Vietnam	3.95–10.80	Ohno et al. (1995)
Cuba	Up to 10	Areces (1995)
Okha (India)	2.45–7.64	Mairh et al. (1995)
Vietnam	3.16–10.8	Ohno et al. (1996)
Brazil	4.5–8.9	Paula, Pereira and Ohno (1999)
Venezuela	4.4–7.7	Rincones and Rubio (1999)
Philippines	2.3–4.2	Hurtado et al. (2001)
India	0.34–5.5	Eswaran et al. (2002)
Brazil	4.5–8.2	Paula et al. (2002)
Yucatán, México	2.0–7.1	Muñoz et al. (2004)
South-eastern Brazil	5.7	Bulboa and Paula (2005)
Mithapur (India)	3.93–9.80	This study
Okha (India)	3.86–13.98	This study
Beyt Dwaraka (India)	3.64–9.35	This study

The high growth rate and biomass reported in this study was found at a higher seawater temperature (25.4 – 26.9 °C) and this was in agreement with the results reported by Trono and Ohno (1989). Temperature appeared to be the main factor affecting the growth rates of this seaweed as was noticed in the present study and this was also reported by Muñoz et al. (2004) in Yucatan, Mexico, and Ohno et al. (1994) at Shikoku, Japan.

Paula and Pereira (2003) reported a positive correlation between growth rate and temperature with a brown strain of *K. alvarezii* in subtropical waters of Brazil and similar results were obtained at Okha and Beyt Dwarka. The reduction in growth during December and January at the three places studied could be attributed to low seawater temperature (19.3 – 22.8 °C). The seasonal variation in seawater temperature of 28 – 31 °C at Dzilam and 21 – 28 °C at Kaneohe Bay, Hawaii (Glenn & Doty 1990), 17 – 33 °C Ubatuba Bay, Brazil (Paula & Pereira 2003), 19.9 – 26.5 °C at Mithapur, 19.3 – 26.9 °C at Okha and 19.8 – 25.4 °C at Bet Dwarka may explain the difference in growth between subtropical and tropical locations.

It is believed that temperature, light intensity and nutrients are the most important parameters governing the growth of *Kappaphycus* and a low correlation was observed between growth and environmental factors in this seaweed (Glenn & Doty1990) and this was confirmed in the present study where a low correlation was found between growth rate and salinity at Okha, between growth rate, phosphate and temperature at Mithapur and between growth rate, phosphate and salinity at Beyt Dwaraka. The highest growth rates were found during October/March in the present study at the all the places while the same was reported during April to December in Fiji (Prakash1990) and September to February in Philippines (Hurtado et al. 2001). No grazing and disease were found at the experimental sites. The introduction of this seaweed into Indian waters had no deleterious effect on the natural biota based on the experimental evidence, which has also been reported elsewhere by McHugh (2001) and Paula et al. (2002). However, a quarantine protocol was adopted and the seaweed was domesticated under controlled conditions in order to get the seaweed acclimatized to the surrounding environments to minimize the risks of introduction initially at Mandapam, Gulf of Mannar waters, Southeast coast of India, and subsequently at Diu and Okha, Northwest coast of the Indian Peninsula, as advocated by Zemke-White (2002), and Ask and Azanza (2002).

The semi-refined carrageenan obtained from *K. alvarezii* at the three localities was found to range from 42.42 1.89 to 58.36 1.26 and these were comparable with the yields obtained elsewhere for this seaweed. A lower semi-refined carrageenan content (31 – 43%) was reported for this seaweed farmed in the subtropical waters of Sao Paulo State, Brazil (Hayashi, Paula & Chow 2007). Moderate yields for materials from Vietnam (34.5% to 45.3%) and Indonesia (45%) and the highest yield for material from Philippines (54.6%) were obtained (c.f Ohno et al.1996).The present yield (58.36%) was very much comparable with the one reported for this seaweed from the western Pacific commercial sample (57%) (Hayashi, Oliveira, Lhonneur, Boulenguer, Pereira, Seckendorj, Shimoda, Lefamand, Patrick Valleée & Critchley 2007).

By taking up cultivation on a large scale year round, carrageenan production could be initiated in India, which is currently being completely imported to the tune of 400 tonnes annum1 to meet the internal demand (Sakthivel 2006). The increasing domestic market may boost carrageenan production, indirectly necessitating biomass production of *Kappaphycus alvarezii* through large-scale cultivation.

Because the growth rates obtained at the three study locations were above 3.5%, which is considered to be adequate for commercial farming (Parker 1974; Braud & Perez 1978; Liu & Zhuang 1984; Adnan & Porse 1987; Luxton et al. 1987; Ask & Azanza 2002), commercial cultivation is feasible and may be taken up either by NGOs or Self Help Groups to facilitate development of the rural coastal areas on the one hand and to improve the living standards of the coastal rural people on the other.

References

Adnan, H., & Porse, H. (1987). Culture of *Eucheuma cottonii* and *Eucheuma spinosum* in Indonesia. *Hydrobiologia, 151/152*, 355–358.

Areces, A.J. (1995). Cultivo comercial de carragenofitas del género *Kappaphycus* Doty In: K. Alveal, M.E. Ferrairo, E.C. Oliveira and E. Sar, Editors, Manual de Métodos Ficológicos, Univ. Concepción, Concepción, Chile, pp. 529–550.

Ask, E.I., & Azanza, R.V. (2002). Advances in cultivation technology of commercial Eucheumatoid species: a review with suggestions for future research. *Aquaculture, 206*, 257–277.

Bixler, H.J. (1996). Recent developments in manufacturing and marketing carrageenan. *Hydrobiologia, 326/327*, 35–57.

Braud, J.P., Perez, R., (1978). Farming on a pilot scale of *Eucheuma spinosum* (Florideophyceae) in Djibouti waters. In Jenson, A., Stein, J. (Eds.), Proceedings of the 9th International Seaweed Symposium. Science Press, Princeton. pp. 533–539.

Bulboa, C.R., & Paula, E.J.D. (2005). Introduction of non-native species of *Kappaphycus* (Rhodophyta, Gigartinales) in subtropical waters: comparative analysis of growth rates of *Kappaphycus alvarezii* and *Kappaphycus striatumin* vitro and in the sea in south-eastern Brazil. *Phycological Research, 53*, 183–188.

Christopher, S., & Michael, G. (1997). Method for extracting semi-refined carrageenan from seaweed. United States Patent No.5801240.

Dawes, C.J., Lluisma, A.O., & Trono, G.C. (1994). Laboratory and field growth studies of commercial strains of *Eucheuma denticulatum* and *Kappaphycus alvarezii* in the Philippines. *Journal of Applied Phycology, 6*, 21–24.

De Reviers, B. (1989). Réalization dúne ferme de culture industrielle de Eucheuma aux Maldives. *Oceanis, 15*, 749–752.

Doty, M.S. (1973). Farming the red seaweed, Eucheuma, for carrageenan. *Micronesia, 9*, 59-73.

Doty, M.S., & Alvarez, V.B. (1975) Status, problems, advances and economics of *Eucheuma* farms. *Journal of Marine Technological Society, 9*, 30–35.

Doty, M.S., & Alvarez, V.B. (1981). Eucheuma farm productivity. In: International Seaweed Symposium (ed. by G.E. Fogg & W.E. Jones), Vol. 8, pp. 688–691. The Marine Science Laboratories, Menai Bridge.

Eswaran, K., Ghosh P.K., & Mairh, O.P. (2002). Experimental field cultivation of *Kappaphycus alvarezii* (Doty) Doty.ex. P.Silva at Mandapam region. *Seaweed Research and Utilization, 24*, 67–72.

Glenn, E.P., & Doty, M.S. (1981). Photosynthesis and respiration of the tropical red seaweeds, Eucheuma striatum (tambalang and elkhorn varieties) and E. denticulatum. *Aquatic Botany, 10*, 353–364.

Glenn, E.P., & Doty M.S. (1990). Growth of the seaweeds *Kappaphycus alvarezii, Kstriatumand Eucheuma denticulatum* as affected by environment in Hawaii. *Aquaculture, 84*, 245–255.

Glenn, E.P., & Doty, M.S. (1992). Water motion affects the growth rates of *Kappaphycus alvarezii* and related red seaweeds. *Aquaculture, 108,* 233–246.

Hayashi, L., Paula, E.J.D., & Chow F. (2007). Growth rate and carrageenan analyses in four strains of *Kappaphycus alvarezii* (Rhodophyta, Gigartinales) farmed in the subtropical waters of Sao Paulo State. *Brazil Journal of Applied Phycology, 19,* 393–399.

Hayashi, L., Oliveira, E.C., Lhonneur, G.B., Boulenguer, P., Pereira, R.T.L., Seckendorff, R.V., Shimoda, V.T., Lefamand, A., Vallée P., & Critchley A.T. (2007). The effects of selected cultivation conditions on the carrageenan characteristics of *Kappaphycus alvarezii* (Rhodophyta, Solieriaceae) in Ubatuba Bay, Sao Paulo. *Brazil Journal of Applied Phycology, 19,* 505–511.

Hurtado, A.Q. (1995). Carrageenan properties and proximate composition of three morphotypes of *Kappaphycus alvarezii* Doty (Gigartinales Rhodophyta) grown at two depths. *Botanica Marina, 38,* 215–219.

Hurtado, A.Q., Agbayani, R.F., Sanares, R., & Castro-Mallare M.T.R. (2001). The seasonality and economic feasibility of cultivating *Kappaphycus alvarezii* in Panagatan Cays, Caluya, Antique Philippines. *Aquaculture, 199,* 295–310.

Hurtado-Ponce, A.Q. (1992). Cage culture of *Kappaphycus alvarezii* var. tambalang (Gigartinales, Rhodophyceae). *Journal of Applied Phycology, 4,* 311–313.

Li R., Li J. &Wu C. (1990). Effect of ammoniumon growth and carrageenan content in *Kappapycus alvarezii* (Gigartinales, Rhodophyta). *Hydrobiologia, 204/205,* 499–503.

Lirasan, T., & Twide P. (1993). Farming Eucheuma in Zanzibar, Tanzania. *Hydrobiologia, 260/261,* 353–355.

Liu, S., & Zhuang, P. (1984). The commercial cultivation of Eucheuma in China. *Proceeding of International Seaweed Symposium, 11,* 243–245.

Luxton, D.M. (1993). Aspects of the farming and processing of *Kappaphucus* and *Eucheuma* in Indonesia. *Hydrobiologia, 260/261,* 365–371.

Luxton, D.M., & Luxton P.M. (1999). Development of commercial *Kappaphycus* production in the Line Islands, Central Pacific. *Hydrobiologia, 398/399,* 477–486.

Luxton L.M., Robertson M. & Kindley M.J. (1987) Farming of Eucheuma in the South Pacific Islands of Fiji. *Hydrobiologia, 151/152,* 359–362.

Mairh, O.P., Zodape, S.T., Tewari, A., & Rajyaguru, M.R. (1995). Culture of Marine Alga *Kappaphycus striatum* (Schmitz) Doty on the Suarashtra region, Westcoast of India. *Indian Journal of Marine Science, 24,* 24–31.

McHugh, D.J. (2001). Prospects for seaweed production in developing countries. FAO Fisheries Circular No. 968 FIIU/C968, 28pp.

Mollion, J., & Braud, J.P. (1993). A *Eucheuma* (Solieriaceae, Rhodophyta) cultivation test on the south-west coast of Madagascar. *Hydrobiologia, 260/261,* 373–378.

Muñoz, J., Freile-Pelegrín, Y., & Robledo, D. (2004). Mariculture of *Kappaphycus alvarezii* (Rhodophyta, Solieriaceae) color strains in tropical waters of Yucatán, México. *Aquaculture, 239,* 161–177.

OhnoM., Largo, D.B. & Ikurnoto, T. (1994). Growth rate, carrageenan yield and gel properties of culture kappa carrageenan producing red alga *Kappaphycus alvarezii* (Doty) Doty in the subtropical waters of Shikoku, Japan. *Journal of Applied Phycology, 6,* 1–5.

Ohno, M., Nang, H.O., Dinh, N.H., & Triet, V.D. (1995). On the growth of cultivated *Kappaphycus alvarezii* in Vietnam. *Japanese Journal of Phycology (Sorui)*, 43, 19–22.

Ohno, M., Nang, H.O., & Hirase, S. (1996). Cultivation and carrageenan yield and quality of *Kappaphycus alvarezii* in the waters of Vietnam. *Journal Applied Phycology, 8,* 43–437.

Parker, H.S. (1974). The culture of the red algal genus Eucheuma in the Philippines. *Aquaculture, 3,* 425–439.

Paula, E.J., & Pereira, R.T.L. (2003). Factors affecting growth rates of *Kappaphycus alvarezii* (Doty) Doty ex P. Silva (Rhodophyta, Solieriaceae) in subtropical waters of Sao Paulo State, Brazil. In: Proceedings of the XVII International Seaweed Symposium (ed. by A.R.O. Chapman, R.J. Anderson, V.Vreel and & I. Davison), pp. 381–388. Oxford University Press, NewYork.

Paula, E.J., Pereira, R.T.L., & Ohno, M. (1999). Strain selection in *Kappapycus alvarezii* var. *alvarezii* (Solieriaceae, Rhodophyta) using tetraspore progeny. *Journal of Applied Phycology, 11,* 111–121.

Paula, E.J., Pereira, R.T.L., & Ohno, M. (2002). Growth rate of carrgeenophyte *Kappaphycus alvarezii* (Rhodophyta, Girtinales) introduced in subtropical water of Sao Paulo State, Brazil. *Phycological Research, 50,* 1–9.

Prakash, J. (1990). Fiji. In Proceedings of the Regional Workshop on Seaweed Culture and Marketing. South Pacific Aquaculture Development Project, Food and Agriculture Organization of the United Nations, Suva, Fiji, 14–17 November 1989 (ed. by T. Adams & R. Foscarini), pp 1–9. FAO, Rome, Italy.

Qian, P.Y., Wu, C.Y., Wu, M., & Xie Y.K. (1996). Integrated cultivation of the red alga *Kappaphycus alvarezii* and the pearl oyster *Pinctada martensi*. *Aquaculture, 147,* 21–35.

Rincones, R.E., & Rubio, J. (1999). Introduction and commercial cultivation of the red alga *Eucheuma* inVenezuela for the production of phycocolloids. *World Aquaculture, 30,* 57–61.

Sakthivel, M. (2006). *Kappaphycus* seaweed cultivation: economics. *Fishing Chimes*, 26, 19–24.

Strickland, J.D.H., & Parsons, T.R. (1972). Apractical handbook of seawater analysis. *Fisheries Research Board of Canadian Bulletin, 1,* 167–311.

Trono, G.C., & Ohno, M. (1989). Seasonality in the biomass production of the Eucheuma strains in Northern Bohol, Philippines. In: Scientific Survey of Marine Algae and their Resources in the Philippine Islands (ed. by I. Umezaki), pp. 71–80. Monbushio International Scientific Research Program, Japan.

Zemke-White, W.L. (2002). Assessment of the current knowledge on the environmental impacts of seaweed farming in the tropics. Proceedings of the Asia-Pacific Conference on Marine Science and Technology, Kuala Lumpur, Malaysia, 88 pp.

1. Introduction

Climate change and over-exploitation of natural resources impose challenges to various coastal species, including industrially important seaweeds. Since natural seaweed beds are insufficient in coping with the demand of the global seaweed industries, aquaculture seems to be the sole alternative to generate raw materials and avoid subjugation of natural beds (Góes and Reis 2011). Seaweed farming helps meet the industry market; it prevents over-exploitation of valuable ecosystems. Besides it has socio-economic advantages (creating self-employment opportunities as a means of livelihood) (Periyasamy et al. 2014a), and, is helps climate protection (Hayashi et al. 2010). Seaweed farming do not involve nursery-rearing of the seedlings and do not demand expensive infrastructural support or expertise, and, this could be the reason for its immense popularity in developing countries (Bast 2014). Researchers encourage seaweed cultivation keeping in mind that seaweeds once cultivated could also be put to several potential applications including biofuel production (Sivasubramanian 2011; Meinita et al 2011; Khambhaty et al. 2012; Hargreaves et al. 2013).

Amongst the various seaweeds cultivated (e.g. Porphyra, Laminaria, Macrocystis, Gracilaria, Undaria and Eucheuma), *Kappaphycus alvarezii*, a tropical red seaweed is of enormous economic importance (Thirumaran and Anantharaman 2009; Msuya et al. 2014). It has a worldwide industrial demand for its cell wall polysaccharide - carrageenan (a family of linear sulphated polysaccharides) which is utilized as a gelling, viscosity-enhancing, texture-modifying, and cell-immobilizing agent in various food, pharmaceutical, industrial, and biotechnological applications (Bixler 1996; Ask and Azanza 2002; Villanueva et al. 2011). Its remarkably fast growth rate (with a potential to double in size every 15 to 30 days) and the ease-in-handling, prove to be assets, making *Kappaphycus alvarezii* the seaweed of choice for farming purposes. In 2009, *K. alvarezii* production was 1,755 tons, which was almost worth 203 million US$ (Hayashi and Reis 2012). But Bixler and Porse (2011) pointed out that the entire global harvest of *Kappaphycus* (i.e. 183,000 t dry) came from cultivation alone.

Cultivation of *K. alvarezii* originated in southern Mindanao, Philippines; subsequently, this seaweed was introduced globally to several maritime countries for experimental and commercial cultivation as a sustainable alternate livelihood for coastal villagers. In fact, *K. alvarezii* along with *K. striatum* (Schmitz) Doty from the Philippines have been introduced into more than 20 tropical countries as part of coastal management (Thirumaran and Anantharaman 2009). *K. alvarezii* is one of the most popular seaweed, cultivated in Hawaii, India, Japan, Kenya, Mexico, Kiripati, Micronesia, Indonesia, Vietnam, Fiji, Maldives, Dzilam, China, Tanzania, Cuba, Brazil, and Venezuela (Góes and Reis 2011; Subba Rao et al. 2008; Hurtado-Ponce et al. 2001). Philippines, Indonesia, Malaysia (Sabah), Fiji and Tanzania comprise the commercial producers and provide a substantial continuous flow of this seaweed. Commercial cultivation of *K. alvarezii* was developed jointly by Marine Colloids Corporation (purchased by FMC Corporation in 1977, now part of BioPolymer) and Maxwell Doty, University of Hawaii (Subba Rao et al. 2008; Bindu and Levine 2011a). Experimental cultivation of this seaweed has been successfully reported in Madagascar (Mollion and Braud 1993), Vietnam (Ohno et al. 1996), China (Qian et al. 1996) and India (Periyasamy et al. 2014a,b; Athithan,

2014; Thirumaran and Anantharaman 2009). The rising demand of this seaweed, and the increasing profit that could be gained from its product carrageenan has been discussed globally (Bixler and Porse 2011), including several reports from India (Krishnamurty 2005; Mishra et al. 2006; Johnson and Gopakumar 2011; Periyasamy et al. 2014b).

Mairh et al. (1995) pioneered the commercial introduction of eucheumatoid *Kappaphycus striatum* to the Indian waters during 1989, followed by which tide pool cultivation experiments with this seaweed were undertaken at Okha, Northwest coast of India (Mairh et al. 1995). Apart from reports demonstrating feasible cultivation of *K. alvarezii* (Doty) Doty on the Southeast coast of India and Northwest coast (Eswaran et al. 2002; Subba Rao et al. 2008), several findings also detail the acceptance and ease of *Kappaphycus* cultivation in the Indian waters (Thirumaran and Anantharaman 2009; Bindu 2011b). Bindu (2011b) and Periyasamy et al. (2014a) reported cultivation and processing of *K. alvarezii*, to help social upliftment and empowerment of coastal residents in India. The utilization of seaweed cultivation technologies and their dissemination have been long discussed (Thirumaran and Anantharaman 2009; Johnson and Gopakumar 2011). Reports explicating experimental field cultivation of economically important seaweeds in different maritime states of India, their vegetative propagation, and the various culture techniques, are available (Thirumaran and Anantharaman 2009). In the Indian context, commendable strides in technology developments have been initiated by CSMCRI (a CSIR lab, Bhavnagar, India), Gujarat State Biotechnology Mission (GSBTM), Government of Gujarat the Department of Science & Technology, and, Aquagri Processing Private Limited.

The tropical climate, ecological conditions and surrounding ocean waters project the Indians waters to be suitable for farming *Kappaphycus* species (Bindu 2011b); warm, nutrient-enriched seawater, high light levels and high degree of water motion are known to facilitate successful cultivation of *Kappaphycus* species (Muñoz et al. 2004; Bast 2014). Despite possessing environmental conditions that favor cultivation of this carragenophyte, India imports carrageenan (Bindu 2011b). Large-scale commercial cultivation of this species is yet to gain momentum throughout the country. These reports emphasize upon the importance of carrying out *Kappaphycus* cultivation in India. The National Academy of Agriculture Sciences India has declared commercial cultivation of *Kappaphycus alvarezii* as ecologically safe (National Academy of Agriculture Sciences 2003). Commercial cultivation of *K. alvarezii* could not only improve the economic status of the fisher folk but would also help in establishing a new course to the export corridor. There seems to be an overall lack of awareness regarding the know-how of cultivation of this species, its benefits and feasibility with the Indian context. Thus, it is essential to explicitly detail various technicalities of *Kappaphycus alvarezii* cultivation, including specifying good cultivation practices that would lead to better yield and product quality. In addition, it also helps economize the investment cost and manpower involved. Information regarding the growth of *K. alvarezii* cultivated in commercial farms, the carrageenan yield and quality, and recognition of the effect of the environmental factors on these variables, are fundamental for management activities (Ask and Azanza 2002; Hayashi et al. 2007; Reis et al. 2007). This helps subsidize the establishment cost, and aids efficient management activities in the farms. Santelices (1999) reported that the productivity of an established farm depends on the management efficiency. Efforts have been carried out worldwide to evaluate the feasibility of producing biomass for the carrageenan industry. For instance, Neish (2008) has explicitly detailed good agronomy practices to improve the yield of Kappaphycus and Eucheuma. Thirumaran and Anantharaman (2009) too have detailed several such feasibility studies, including the introduction of *K. alvarezii* in the warm waters of the Caribbean and the Western Atlantic. Despite the availability of few reports (Subba Rao et al. 2008; Mairh et al. 1995; Eswaran et al. 2002; Dash et al. 2009; Thirumaran and Anantharaman 2009; Bindu 2011b; Periyasamy et al., 2014a, b) it is extremely essential to conduct more exquisite baseline research on this aspect with an Indian perspective. In this regard, this study demonstrates the effect of various environmental parameters, culturing techniques and culturing period on the growth rate, and, biomass production. Besides it provides an in-depth understanding of the feasibility of cultivation of *K. alvarezii* in the northwest waters of India, thereby advocating seaweed farming as a viable socio-economic option.

2. Materials and methods

Kappaphycus alvarezii (Doty) Doty collected from a cultivation farm at Mandapam coast (Tamil Nadu) India, was nurtured at Diu (West coast) India. Thereafter, 10 kg of the seaweed was primarily collected from the Diu farm for the present study, domesticated at Okha (West coast) India. These domesticated samples were used as seeding material for further study. Cultivation of *K. alvarezii* was carried out from August 2004 to April 2006 at Okha (22° 28.656′ N and 69° 04.015′ E) (Fig. 1) using two different methods (i.e. raft and net bag).

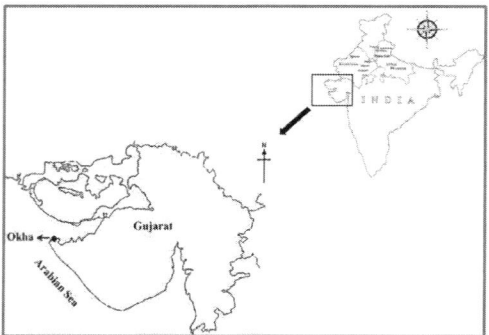

Fig. 1. Map showing the location of experimental sites on the Northwest coast, India

2.1. Raft cultivation

The seeding material for raft cultivation comprised of 100g young fragments of *K. alvarezii*. The size of the raft (2.5 m × 2.5 m) was chosen to suit the experimental site. The raft contained 16 monolines of 3 mm polypropylene rope (2.5 m long) at 15 cm intervals; each line consisted of 16 seedlings (each 100 g fresh weight) at 15 cm intervals. The seedlings were tangled onto the monoline using the tie-tie method (Doty and Alvarez 1975). Each raft contained approximately 25.6 kg seed material. The rafts were anchored with the help of a polypropylene rope tied onto concrete blocks. Five such rafts were anchored into the sea at experimental sites that were not generally exposed during low tides. After 15, 30 and 45 days of growth, five plants were randomly collected from each raft followed by which the Daily Growth Rate (DGR) and biomass were evaluated. The entire biomass was harvested after 45 days to determine the crop yield per raft.

2.2. Net bag cultivation

For the net bag studies, monolines (comprised a 10 mm polypropylene rope) were fastened at two ends of the experimental site and kept floating with the help of floaters. The seed material, i.e. 100 g *K. alvarezii*, was planted onto the monoline with the help of soft plastic fabric using the tie-tie method (Doty and Alvarez 1975). Three such monolines were planted; each monoline contained 20 seedlings at 30 cm interval. Each seedling was enclosed in an old fish net bag (60 × 40 cm size) made up of 2 mm high density polyethylene (HDPE) thread, having a mesh size of 1.5 cm. Five plants were collected randomly from each monoline on a fortnightly basis up to 45 days. Growth in terms of Daily Growth Rate (DGR) and biomass was evaluated after 15, 30 and 45 days; the entire biomass was harvested after 45 days to determine the crop yield line^{-1}.

2.3. Determination of daily growth rate, biomass and crop yield

The Daily Growth Rate DGR (%) was calculated using the formula of (Dawes et al. 1994). Biomass yield was determined by gravimetrically weighing the fresh harvested plant material.

$$DGR \% = In (W_f / W_o) / t \ X \ 100$$

Where W_f is the final fresh weight (g) at t day, W_o is the initial fresh weight (g), t is the number of culture days. The quantity of fresh biomass obtained per raft was determined and presented as the crop yield (kg FW raft-1) in case of raft cultivation; on the other hand, in case of the net bag method, the quantity of fresh biomass obtained per line was determined and presented as the crop yield (kg FW line^{-1}).

2.4. Environmental parameters

Environmental parameters were determined at the farming site during the period of experimental cultivation (September 2004 to April 2006). All determinations, including water sample collections, were conducted at a depth of 30 cm i.e. the same depth as the cultured seaweeds. Seawater temperature was recorded (with a thermometer) daily throughout the study period and the mean monthly temperature was obtained. Salinity was determined fortnightly and the average value was recorded (using conventional laboratory type Baume meter). Nutrient analysis of the seawater samples were carried out fortnightly. The dissolved inorganic phosphate (P–PO$_4$) and dissolved inorganic nitrate (N–NO$_3$) were determined according to the method described by Strickland & Parsons (1972) followed by which average monthly values were calculated.

2.5. Statistical analysis

Analysis of variance (ANOVA) was conducted to determine the differences in seasonal biomass yield, DGR and duration of culture period i.e. 15, 30 and 45 days. The biomass (fresh) and growth rates were expressed in terms of mean ± standard deviation. Multiple comparison tests with the least significance difference (LSD) were performed to determine significant differences.

3. Results

During the period of study (September 2004-April 2006), mean salinity and seawater temperature were 34.20 ± 1.84‰ and 26.55 ± 1.76°C respectively (Fig. 2).

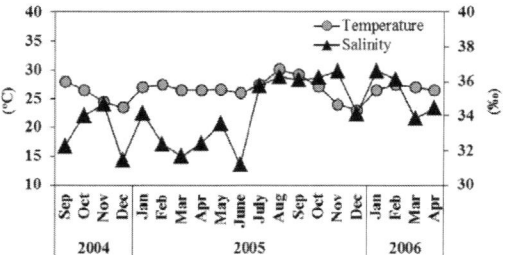

Fig. 2. Monthly variation in surface seawater temperature and salinity recorded at the cultivation

The average nitrate and phosphate content recorded were 17.46 ± 7.11 μM L^{-1} and 1.33 ± 0.76 μM L^{-1} respectively (Fig. 3)

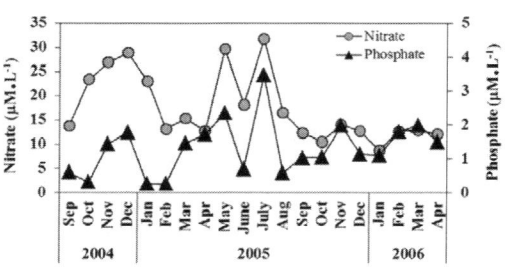

Fig. 3. Monthly variation in phosphate and nitrate recorded at the cultivation

3.1. Daily Growth Rate (DGR) and Biomass

3.1.1. Raft cultivation

After fifteen days of growth, maximum DGRs were recorded during August 2004 and August 2005 (i.e. 9.49 ± 0.07 and 9.26 ± 0.12% respectively, Fig. 4a). In addition, high biomass yield per plant (i.e. 415.0 ± 4.24 and 401.0 ± 7.07 g FW respectively) were also recorded during these months (Fig. 5a). Low growth rates and complying biomass yield per plant were observed in November 2004 (4.51 ± 0.43%; 197.0 ± 12.73 g FW respectively) and October 2005 (4.24 ± 0.05%; 189.0 ± 1.41 g FW respectively). These values significantly differed from the values obtained during peak growth period ($p < 0.05$).

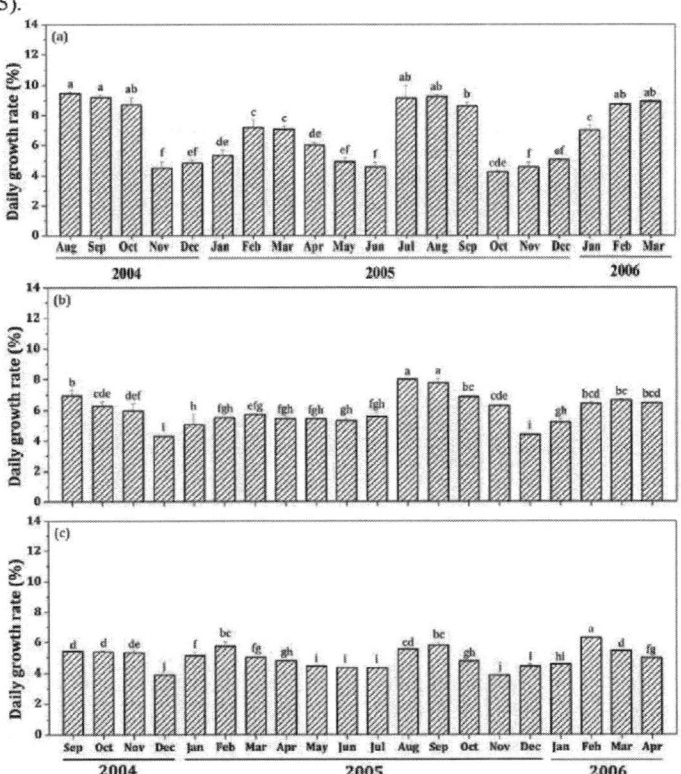

Fig. 4. Effect of duration of culture period on DGR (%) of *K. alvarezii* (raft method) (a) 15 days, (b) 30 days and (c) 45 days. Error bars are standard deviations. Different small letters over bars denote significant difference at $p < 0.05$

In case of the plants harvested after 30 days of growth, notable DGRs of 7.99 ± 0.02 and 7.78 ± 0.27% were observed during August 2005 and September 2005 (Fig. 4b) with maximum biomass yields 1100.0 ± 5.66 g FW and 1033.5 ± 84.15 g FW respectively (Fig. 5b). Minimum daily growth rates were recorded in the December 2004 (4.32 ± 0.01) and December 2005 (4.42 ± 0.03 %) respectively with corresponding biomass values (365.0 ± 1.41 g FW and 376.0 ± 2.83 g FW). These values significantly differed from the values obtained during peak growth period ($p < 0.05$). Highest biomass yield per plant was recorded in September 2004 (805.0 ± 94.75 g FW), which gradually declined until December 2004. Subsequently there was a plodding increase in the biomass yield per plant until August 2005.

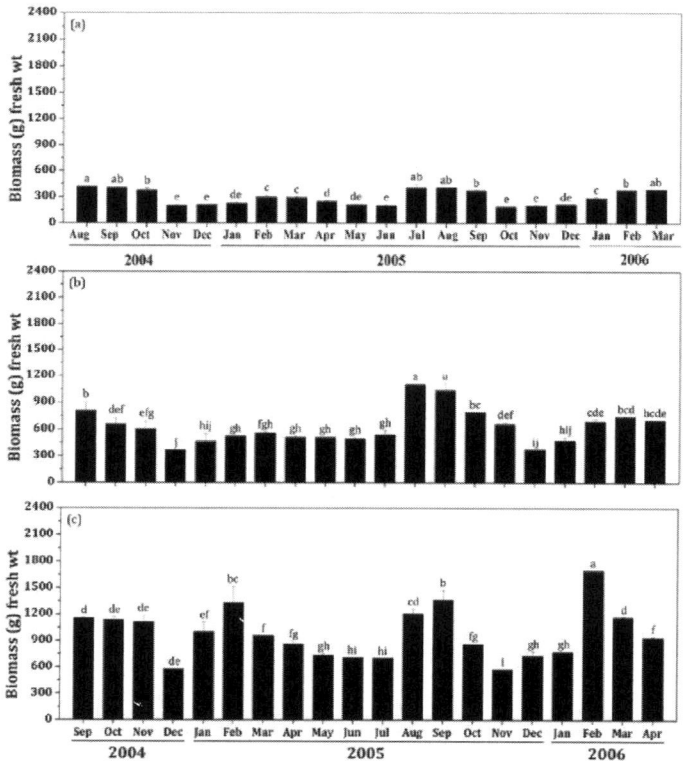

Fig. 5. Effect of duration of culture period on biomass yield plant^{-1} (g FW) of *K. alvarezii* (raft method) (a) 15 days, (b) 30 days and (c) 45 days. Error bars are standard deviations. Different small letters over bars denote significant difference at $p < 0.05$

The forty-five days grown samples demonstrated remarkably good DGRs (5.43 ± 0.02 and $5.80 \pm 0.18\%$ respectively) in the month of September during the two consecutive years (Fig. 4c) i.e. September 2004 and September 2005, with biomass yields of 1153.0 ± 9.90 g FW and 1363.5 ± 108.19 g FW respectively (Fig. 5c). However, in February 2006, high DGR ($6.30 \pm 0.00\%$) with highest biomass of 1699.0 ± 2.83 g FW were recorded; this was the maximum value recorded during the entire study period. On the other hand, low DGRs were evidenced in December 2004 ($3.88 \pm 0.01\%$) and November 2005 ($3.86 \pm 0.02\%$) accompanied by low biomass values (i.e. 572.0 ± 2.83 g FW and 569.0 ± 4.24 g FW respectively), which significantly differed from the values obtained during peak growth period ($p < 0.05$). A decrease in biomass was observed from September 2004 to December 2004, after which an increase was observed with a maximum in September 2005. This was similar to observations obtained in case of the 15 and 30 day samples.

During the entire cultivation period, the mean DGRs of 6.87 ± 0.28, 5.99 ± 0.18 and $4.97 \pm 0.08\%$ were recorded respectively in case of the plants harvested after 15, 30 and 45 days of growth, with respective biomass values per plant as 292.43 ± 12.06, 629.58 ± 32.56 and 977.96 ± 36.59 g FW in raft cultivation. Moreover, in the samples harvested after 15 days as well as 45 days apparent biomass increases with decrease in DGRs were evidenced. Notable quantities of healthy seaweed biomass were obtained from each raft, thereby emphasizing on the appropriateness of raft method for *K. alvarezii* cultivation. The mean crop yield during the study period was 163.60 ± 4.81 kg FW raft^{-1}. Minimum crop yields were observed during December 2004 (96.0 ± 5.66 kg FW raft^{-1}) and November 2005 (81.0 ± 1.41 kg FW raft^{-1}). The highest crop yield of 224.5 ± 3.54 kg FW raft^{-1} was obtained in August 2005.

3.1.2. Net bag cultivation

In case of net bag cultivated samples harvested after fifteen days, highest DGR values were evidenced in June 2005 ($13.18 \pm 0.04\%$). Commendable DGRs with corresponding biomass yields were obtained during February 2005 ($9.67 \pm 1.30\%$; 431.0 ± 83.44 g FW respectively) and July 2005 (473.5 ± 67.18 g FW; $10.30 \pm 0.94\%$ respectively) (Fig. 6 & 7). A fluctuation in biomass per plant value was evidenced during 2005. Nevertheless, low daily growth rates were recorded in November 2004 ($3.68 \pm 0.71\%$) and December 2005 ($6.63 \pm 0.35\%$), with correspondingly low biomass values (175.0 ± 18.39 g FW and 271.0 ± 14.14 g FW respectively). The biomass and DGRs obtained during the peak period of growth were significantly different ($P < 0.05$).

In case of the net bag samples harvested after 30 days, highest growth rates were documented in September 2004 ($7.54 \pm 0.08\%$ with corresponding biomass as 966.0 ± 25.46 g FW) and February 2005 ($7.48 \pm 0.08\%$; biomass 945.0 ± 21.21 g FW). The minimum DGRs (3.76 ± 0.06 and $3.69 \pm 0.11\%$ respectively) were recorded in December 2004 and May 2005 when the biomass of the plant was about 311.0 ± 7.07 g FW and 303.0 ± 9.90 g FW respectively. Here, the biomass and DGRs obtained during the peak period of growth were significantly different ($P <$ 0.05). A relatively consistent growth was observed from March 2005 to September 2005.

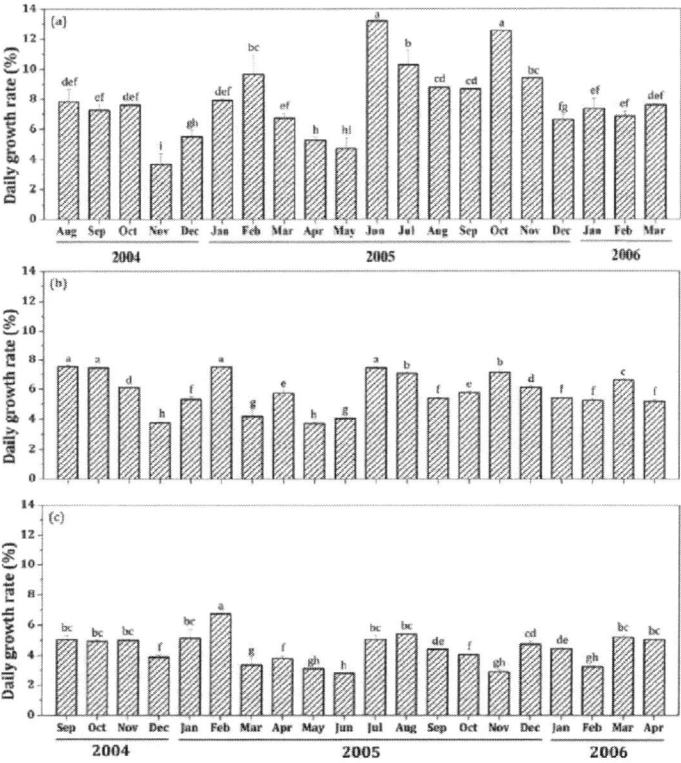

Fig. 6. Effect of duration of culture period DGR (%) of *K. alvarezii* (net bag) (a) 15 days, (b) 30 days and (c) 45 days. Error bars are standard deviations. Different small letters over bars denote significant difference at $p < 0.05$

Forty-five days net-bag grown samples (Fig. 6c) showed commendable growth rates in February 2005 ($6.76 \pm 0.02\%$) and August 2005 ($5.44 \pm 0.02\%$) with respective biomass values per plant as 2101.0 ± 15.56 g FW and 1155.0 ± 9.90 g FW (Fig. 7c). In this case, minimum DGRs were evidenced in June 2005 and November 2005 (2.78 ± 0.02 and $2.89 \pm 0.15\%$ respectively with low biomass values as 350.0 ± 2.83 g FW and 375.0 ± 15.56 g FW correspondingly). Moreover, the biomass and DGRs recorded during the peak period of growth were significantly different ($P <$

0.05). Relatively higher biomass (>900 g FW plant-1) were obtained in September 2004, October 2004, November 2004, January 2005, July 2005, March 2006 and April 2006.

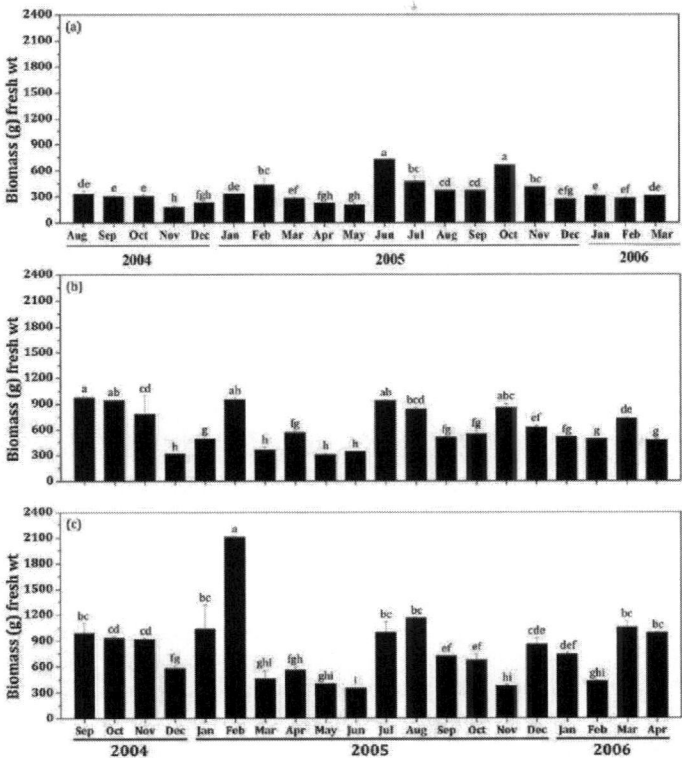

Fig. 7. Effect of duration of culture period on biomass yield plant^{-1} (g FW) of *K. alvarezii* (net bag) (a) 15 days, (b) 30 days and (c) 45 days. Error bars are standard deviations. Different small letters over bars denote significant difference at $p < 0.05$

During the entire cultivation period, samples harvested after 15, 30 and 45 days of net bag cultivation had mean DGRs of 7.88 ± 0.38, 5.83 ± 0.10 and $4.42 \pm 0.15\%$ respectively, with corresponding biomass of 348.98 ± 19.48, 623.26 ± 29.95 and 813.73 ± 52.64 g FW respectively per plant. As evidenced in case of raft cultivation, the 15 to 45 days net bag samples also demonstrated an increase in biomass with decrease in DGRs. In the net bag cultivation, the average crop yields recorded during the study period were highest in February 2005 (47.5 ± 0.71 kg FW line^{-1}) and March 2006 (51.5 ± 3.54 kg FW line^{-1}). Minimum crop yields were evidenced in April 2005 (8.5 ± 0.71 kg FW line^{-1}) and November 2006 (6.5 ± 0.71 kg FW line^{-1}). The mean crop yield obtained during the study period was 20.78 ± 1.17 kg FW line^{-1}. Overall, substantial amount of seaweed biomass could be acquired per line using the net bag method.

3.1.3. Comparison between the two cultivation methods

In order to have a vivid idea, plants grown using both the cultivation techniques and harvested after forty-five days of growth were compared. The samples obtained using raft cultivation method had DGRs ranging from 3.86 ± 0.02 to 6.30 ± 0.00 % (mean 4.97 ± 0.08 %) and the biomass per plant varied from 569.0 ± 4.24 to 1699.0 ± 2.83 g FW (mean 977.96 ± 36.59 g FW). On the other hand, the DGRs of the net bag cultivated samples ranged from 2.78 ± 0.02 to 5.44 ± 0.02 % (mean 4.42 ± 0.14 %) with respective biomass per plant values ranging from 350.0 ± 2.82 to 1155.0 ± 9.90 g FW (mean 813.73 ± 52.64 g FW), with an exception of February 2005 (DGR 6.76 ± 0.02 %, biomass 2101.0 ± 15.56 g FW). These values demonstrate the superiority of the raft cultivation method in the biomass and DGR perspective, i.e. not only was the mean DGR value obtained using raft cultivation slightly higher than the net bag technique, but also the average biomass per plant obtained in case of the raft cultivation was also much higher than the net bag method. On the whole, the raft method of cultivation yielded higher mean biomass values than the net bag method. Nevertheless, commendable growth rates and biomass were obtained using net bag method too. However, loss of biomaterial was slightly higher in the raft method rather than the net bag cultivation. Therefore, looking into the pros and cons, it could be stated that both the methods were equally feasible for cultivation of *K. alvarezii*.

This study largely indicated that, in both the cultivation methods consistent growth (high biomass) were obtained from August to November, and also from February to April, in this particular site located in western India. Remarkably, the month of February proved to be encouraging in terms of DGR, as in both the methods the best DGR values were accomplished in the month of February i.e. February 2006 for raft and February 2005 for net bag method. This could be attributed to the suitable growth environment prevalent during that time of the year. However, the months of May, June, July and December were not conducive enough, as lowest biomass values were recorded during these months. This could either be due to the rough sea conditions prevailing during this tenure, or the non-conducive environment. Particularly, the minimum biomass values charted in the month of December could be attributed to low seawater temperature ($< 25°C$).

4. Discussion
4.1. Environmental parameters

In an attempt to emphasize on the feasibility of *Kappaphycus alvarezii* cultivation, in this investigation *Kappaphycus alvarezii* was grown in subtropical waters of Northwest coast Okha of the Indian Peninsula during September 2004-April 2006. Apart from comparing the feasibility of two different cultivation methods, viz. raft and net bag cultivation, this study also summarizes various environmental parameters effecting *Kappaphycus* cultivation. During the investigation, the salinity of the seawater was approximately 34.20 ‰ and the temperature mean was about 26.5°C. The seawater was enriched with nitrate and phosphate that fluctuated seasonally during the study. Kotiya et al. (2011) reported analogous values for salinity, temperature, nitrate content and phosphate content in a similar locale. The values obtained in our study are also comparable with the discrete report of Góes and Reis (2011). They narrate temperature ranging from 25 to 28 °C (above 20 °C), and salinity from 30 to 40 ‰ to be optimal, but strong wave action and the wind direction are considered undesirable (Góes and Reis 2011). In the present study, temperature played a pivotal role in regulating growth rates of this seaweed. Precisely, high growth rate and biomass were found at a higher seawater temperature (23.0-30.1 °C), which was in compliance with the report of Trono and Ohno (1989). Reports from other regions such as Mexico (Muñoz et al. 2004), Japan (Ohno et al. 1994) and Brazil (Paula and Pereira 2003), also affirm the role of temperature in cultivation of this particular seaweed. In fact, Mairh et al. (1986) undertook field cultivation of E. striatum in Japan over a wide temperature range (14.3–31.2 °C), thereby recording best growth at 21–31°C. They state that the optimum temperature for laboratory plants ranged between 24 and 31°C, but these plants could withstand temperatures ≤17°C (Mairh et al. 1986). In a subsequent study on outdoor chamber cultivation of K. striatum in India, Mairh et al. (1995) reported high growth rates between 23 and 30°C (Ask and Azanza 2002). Nursidi et al. (2013) at Makassar (Indonesia) reveal escalating pH and temperature to cause a decline in growth of Kappaphycus alvarezii; they also report rise in growth with increasing nitrate.

4.2. Growth rates and yields

In the present study, after 45 days of seeding, the maximum growth rates obtained using the raft and net bag method of cultivation were 6.29% and 6.76% respectively, and the overall mean DGRs obtained were 4.97 and 4.42% correspondingly. These values were comparable with Ohno et al. (1994) who detailed growth rates ranging from 3.64 to 4.17 % for the same seaweed in subtropical waters of Shikoku, Japan. However, varying growth rates of Kappaphycus alvarezii have been evidenced in different regions worldwide, for e.g. in the Vietnam waters 3.95 – 10.80 % (Ohno et al. 1995) and 3.16- 10.80 % (Ohno et al. 1996), in subtropical waters of Brazil (3.6 - 8.9 %) (Paula and Pereira 2003), in São Paulo, Brazil (5.2–7.2%) (Hayashi et al. 2007), in Southern Brazil (5.12–4.29%) (Hayashi et al. 2011), in the tropical waters of Yucatan peninsula Mexico (2.0-7.1 %) (Muñoz et al. 2004), in Hawaii (5.06 %) (Glenn and Doty 1990), in Zanzibar, Tanzania (0.5–5.6%) (Msuya and Kyewalyanga 2006) and 1.7 – 6.8% (Msuya 2013), in Philippines (4.5% reported by Gerung and Ohno (1997) and 0.1-8.4 % by Dawes et al. (1994), and in Pangatan Cays, Caluya, Antique, Philippines (3.9-4.2 %) (Hurtado-Ponce et al. 2001). In fact, Glenn and Doty (1990) reported an average growth rate of 2.0 – 3.3 % for brown, green and red strains of this seaweed in Dzilan (Hawaii). Besides, the DGRs obtained in this study were comparable with DGR of Eucheuma cotonii reported by Adnan and Porse (1987), Mollion and Braud (1993), and Luxton et al. (1987), i.e. 2.5-3.5% (Indonesia), 3-4% (Madagascar) and 3.5-3.7 % (Fiji) respectively. Reports encompassing the Indian coast, reveal growth rates ranging from 2.45 to 7.64 % for Kappaphycus striatum (= K. alvarezii) at Okha, Northwest coast (Mairh et al. 1995), and growth rates ranges of 0.34 to 5.5 % at

Mandapam, Southeast coast of India (Eswaran et al. 2002). The growth rates obtained in the present study were comparable to these values. In fact, these growth rates perfectly fall within the range of the former due to the overlapping geographic locale studied. It is essential to note that in the commercial cultivation context, DGR values > 3.5% d^{-1} are considered appreciable (Doty 1987; Glenn and Doty 1990). The DGRs obtained in this investigation comply with this, thereby suggesting the suitability of cultivation of this species on a commercial scale in this region.

The cultivation method and the protocol used are known to influence the growth rate of this seaweed. For example, a mean daily growth rate (DGR) of 4.68% has been reported from commercial cultivation in Philippines using fixed-bottom line method (Trono and Ohno 1989). In Japan, 5% average DGR has been reported for *K. alvarezii* cultured with the aid of raft floating method (Ohno et al. 1994). On the other hand, higher DGRs (4.4–8.9%) were reported when laboratory-nurtured *K. alvarezii* were transplanted to rafts at a field site located at Luzon Island, Philippines (Dawes et al. 1994). Dung and Nhan (2001) report successful experimental cultivation of *K. alvarezii* in Haiquan Lagoon and Cat Ba Bay, Vietnam utilizing floating rafts and hanging-on plastic lines, wherein growth rates obtained in the open sea (5.18–9.82% d^{-1}) were higher than those achieved in sheltered bays (5.25–7.90% d^{-1}). The mean DGR values, obtained in present investigation (using both raft and net bag method), are definitely at par to most of the aforesaid reports. In addition, they also fulfill the commercial cultivation criteria (DGR values > 3.5% d^{-1}). This emphasizes the fact that both the techniques could be potentially utilized for profitable farming of this species.

4.3. Comparing the methods of cultivation

Though numerous reports projecting different methods of cultivation of *Kappaphycus* are available till date, very few of them compare the applicability of each method. In one such report, Krishnan and Kumar (2010) endorse the bamboo raft technique as the most suitable commercially viable method, but they also narrate mono-line cultivation practices to be prone to severe grazing. On the other hand, Góes and Reis (2011) state that the tubular netting was more effective than the tie-tie method, because in the tubular netting cultivation management (time rates used to plant and harvest the seedlings) was 53.6% faster, no seedlings were lost, the cost was lower, and nearly 20% more estimated return were obtained per year, as compared to the tie-tie technique. Hurtado-Ponce et al. (1996) state lower total expenses were incurred in fixed-bottom than in hanging-long line. However in another study at Philippines, Hurtado-Ponce et al. (2001) compare report highest production (in terms of dry weight kg crop^{-1}) of *K. alvarezii* to be obtained using the hanging long line–fixed off-bottom (HL–FB) method, while lowest using the fixed off-bottom (FB) method. However, the hanging long line technique has been suggested as a more lucrative means as a source of positive income, as it had a good ROI (return on investment) and payback period. Hurtado-Ponce et al. (2001) emphasize on best culture techniques to be adopted during certain months of the year to produce the highest yield and income in Philippines. Unlike Hurtado-Ponce et al. (2001), the present study, clearly indicates the suitability of both the net bag and raft cultivation method for all year round use, particularly in the coastal waters of western India.

Most other reports available, individually project the applicability of a single method, i.e. either the use of raft method, or the net bag method, especially in the Indian context. For e.g. few researchers have projected the suitability of raft technique for cultivation of *K. alvarezii* in India (Thirumaran and Anantharaman 2009; Subba Rao et al. 2008; Kotiya et al. 2011; Gunalan et al. 2010), as well as internationally (FAO corporate document repository 2003). Contrarily, Johnson and Gopakumar (2011) stated that initially net-bag technique was practiced in the southern waters of India, however based on the results of more than 120 trials, the bamboo raft technique emerged as the most suitable commercially viable method. Krishnan and Kumar (2010) too outline the timeline of seaweed cultivation in India, clearly stating initiatives of CSMCRI for net bag cultivation. In this context, this study adds a new dimension to this field by proposing net bag cultivation technique to have equal potential as that of raft cultivation. On the other hand, certain reports suggest that raft systems are favorable for cultivation of *Eucheuma* seaplants (Neish 2003) especially in deep sea and in areas with plentiful (that are inappropriate for deep for on/off-bottom systems), however, one also needs to investigate the suitability of net bag method in such zones. More essentially, based on this study, it could be stated that, in case of the net bag technique, less loss of material due to unforeseen weather and wave motion were recorded. Nonetheless, the raft cultivation provided a rich output in form of yield. Apart from being environment-friendly, both the raft and the net-bag techniques described herein are convenient, as the seeding could be carried out at work areas along the shore (rather than inside the sea), thereby eliminating tiresome work in the water. This is not only reduces health risks and safety hazards, but is also economic. Thus in order to develop seaweed farming at the west coast of India, both the net bag and the raft cultivation method, could be adapted with promising production potential, ensuring better income and socio-economic benefits to the coastal dwellers.

With the rise in the total carrageenan market (approx. 3% per year) and the growing acceptance gained by this product as a food additive, carrageenan has squeezed into the packaging of countless products lining the shelves of our grocery and drug stores today. Realizing the value of the carrageenan and perceiving the potential returns projected from mariculture of carrageenan-yielding *Kappaphycus alvarezii*, costal dwellers and small-scale fishermen have acknowledged seaweed as one of the best socio-economic alternate to obtain viable all-year-round income. Today, numerous tropical countries have surged into seaweed cultivation as a sustainable alternative livelihood for coastal villagers, particularly as part of coastal management programs. As seaweed farming is less strenuous, it could be referred as an easier work-option that could be undertaken even on a part-time basis even by coastal women. However, a proper protocol for quarantine and introduction should be followed while cultivation of *K. alvarezii*. Apart from being used as a biofertilizer, serve as a raw material to the bioethanol and carrageenan industries, one should not overlook the huge amount of employment opportunities and income *Kappaphycus* farming would provide. Foreseeing the commercial benefits involved in cultivation of this multifaceted species, and keeping in mind that it could help mitigate the CO_2 in the environment, eliminate eutrophication and cleanse the water bodies by removal of pollutants; it becomes mandatory to evaluate the optimal conditions for its commercial cultivation and to study the role of various environmental parameters and method of cultivation for this species. Additionally, it is necessary to bring out the fact that seaweed cultivation requires a reduced amount of labour. Moreover, a good income could be obtained by undertaking seaweed cultivation, especially *Kappaphycus alvarezii*. It is absolutely necessary to popularize the various methods of cultivation and put-forward the pros and cons of each method, so that the most suitable method could be adapted by the seaweed farmers; it is also essential to state that the suitability of method would definitely vary depending on the prevailing environmental and climatic parameters at each locale.

5. Conclusion

Naturally produced seaweeds are evidently incapable of meeting the enormous demand for phycocolloids worldwide, and this scenario especially holds true in the Indian context. In order to bridge this gap between the demand and supply, efforts have been focused on mariculture practices. However, detailed methodology of cultivation, the optimal environmental conditions, as well as, the prominence of location and climate based cultivation technique, is less discussed. This study on mariculture of an economically important seaweed, *Kappaphycus alvarezii*, serves as a curtain raiser in Indian and tropical scenario. It testifies the feasibility of cultivation of this species using both the raft and the net-bag cultivation methods. Both the techniques could be potentially utilized for profitable farming of this species in this region. Studies on the effect of various environmental parameters, culturing techniques and culturing period on the growth rate, and biomass production, provide an in-depth understanding of the feasibility of cultivation of *K. alvarezii* in the northwest coast of India. Thereby, seaweed farming could be advocated as a viable source of income, providing socio-economic benefits, in this region.

References

Adnan, H., & Porse, H. (1987). Culture of *Eucheuma cottonii* and *Eucheuma spinosum* in Indonesia. *Hydrobiologia, 151/152,* 355–358.

Ask, E.I., & Azanza, R.V. (2002). Advances in cultivation technology of commercial eucheumatoid species: a review with suggestions for future research. *Aquaculture, 206,* 257–277.

Athithan, S. (2014). Growth performance of a Seaweed, *Kappaphycus alvarezii* under lined earthen pond condition in Tharuvaikulam of Thoothukudi coast, South East of India. *Research Journal of Animal, Veterinary and Fishery, 2(1),* 6–10.

Bast, F. (2014). An illustrated review on cultivation and life history of agronomically important seaplants. In: Seaweed: Mineral composition, nutritional and antioxidant benefits and agricultural uses. Vitor Hugo Pomin (eds.) Nova Publishers, New York ISBN: 978-1-63117-571-8, 39-70pp

Bindu, M.S. (2011b). Empowerment of coastal communities in cultivation and processing of *Kappaphycus alvarezii*-a case study at Vizhinjam village, Kerala, India. *Journal of Applied Phycology, 23,* 157–163.

Bindu, M.S., & Levine, I.A. (2011a). The commercial red seaweed *Kappaphycus alvarezii*-an overview on farming and environment. *Journal of Applied Phycology, 23,* 789–796.

Bixler, H.J. (1996). Recent developments in manufacturing and marketing carrageenan. *Hydrobiologia, 326/327,* 35–57.

Bixler, H.J., & Porse, H. (2011). A decade of change in the seaweed hydrocolloids industry. *Journal of Applied Phycology, 23,* 321–335.

Dash, B., Kumar, M.S., & Rao, G.S. (2009). Integration of seaweed (*Kappaphycus alvarezii*) and pearl oyster (*Pinctada fucata*) along with Asian seabass (*Lates calcarifer*) in open sea floating cage off Andhra Pradesh coast. In: Course manual: National training on cage culture of seabass. Imelda, Joseph and Joseph, V Edwin and Susmitha, V (eds.) CMFRI & NFDB, Kochi, 57–59.

Dawes, C.J., Lluisma, A.O., & Trono, G.C. (1994). Laboratory and field growth studies of commercial strains of *Eucheuma denticulatum* and *Kappaphycus alvarezii* in the Philippines. *Journal of Applied Phycology, 6,* 21–24.

Doty, M.S. (1987). The production and use of *Eucheuma*. In: MS Doty; JF Caddy and B Santelices, editors. Case studies of seven commercial seaweed resources, FAO Fisheries Technical Paper 281, Rome. 123–161.

Doty, M.S., & Alvarez, V.B. (1975). Status, problems, advances and economics of *Eucheuma* farms. *Marine Technology Society Journal, 9,* 30–35.

Dung, P.H., & Nhan, P.T. (2001). Some results of study on experimental culture of Elkhorn Sea moss *Kappaphycus alvarezii* in Cat Bay, Hai Phong. Proceedings, Marine Fisheries Research. *Tuyen tap cac cong trinh nghien cuu nghe ca bien, 2,* 537–547.

Eswaran, K., Ghosh, P.K., & Mairh, O.P. (2002). Experimental field cultivation of *Kappaphycus alvarezii* (Doty) Doty. ex. P. Silva at Mandapam region. *Seaweed Research and Utilization, 24,* 67–72.

FAO corporate document repository (2003) A guide to the seaweed industry FAO Fisheries Technical Paper Version: T441 118 pp. ISBN: 9251049580; ISSN: 0429–9345; http://www.fao.org/docrep/006/y4765e/y4765e09.htm [accessed 03.07.14].

Gerung, G.S., & Ohno, M. (1997). Growth rates of *Eucheuma denticulatum* (Burman) Collins et Harvey and *Kappaphycus striatum* (Schmitz) Doty under different conditions in warm waters of Southern Japan. *Journal of Applied Phycology, 9,* 413–415.

Glenn, E.P., & Doty, M.S. (1990). Growth of the seaweeds *Kappaphycus alvarezii, K. striatum* and *Eucheuma denticulatum* as affected by environment in Hawaii. *Aquaculture, 84,* 245-255.

Góes, H.G.D., & Reis, R.P. (2011). An initial comparison of tubular netting versus tie-tie methods of cultivation for *Kappaphycus alvarezii* (Rhodophyta, Solieriaceae) on the south coast of Rio de Janeiro State, Brazil. *Journal of Applied Phycology, 23,* 607–613.

Gunalan, B., Kotiya, A.S., & Jetani, K.L. (2010). Comparison of *Kappaphycus alvarezii* Growth at Two Different Places of Saurashtra Region. *European Journal of Applied Sciences, 2(1),* 10–12.

Hargreaves, P.I., Barcelos, C.A., da Costa, A.C., & Pereira, N Jr. (2013). Production of ethanol 3G from *Kappaphycus alvarezii*: evaluation of different process strategies. *Bioresource Technology, 134,* 257–63.

Hayashi, L., Hurtado, A.Q., Msuya, F.E., Bleicher-Lhonneur, G., & Critchley, A.T. (2010). A review of *Kappaphycus* farming: prospects and constraints. In: Seckbach J, editor. Seaweeds and Their Role in Globally Changing Environments (Cellular Origin, Life in Extreme Habitats and Astrobiology). Netherlands: Springer. p. 251–283.

Hayashi, L., Oliveira, E.C., Bleicher-Lhonneur, G., Boulenguer, P., Pereira, R.T.L., Seckendorff, R., Shimoda, V.T., Leflamand, A., Vallée, P., & Critchley, A.T. (2007). The effects of selected cultivation conditions on the carrageenan characteristics of *Kappaphycus alvarezii* (Rhodophyta, Solieriaceae) in Ubatuba Bay, São Paulo, Brazil. *Journal of Applied Phycology, 19,* 505–511.

Hayashi, L., Santos, A.A., Faria, G.S.M., Nunes, B.G., Souza, M.S., Fonseca, A.L.D., Barreto, P.L.M., Oliveira, E.C., & Bouzon, Z.L. (2011). *Kappaphycus alvarezii* (Rhodophyta, Areschougiaceae) cultivated in subtropical waters in Southern Brazil. *Journal of Applied Phycology, 23,* 337–343.

Hayashi. L., & Reis, R.P. (2012). Cultivation of the red algae *Kappaphycus alvarezii* in Brazil and its pharmacological potential. *Revista Brasileira de Farmacognosia, 22(4),* 748–752.

Hurtado-Ponce, A.Q., Agbayani, R.F., & Chavoso, E.A.J. (1996). Economics of cultivating *Kappaphycus alvarezii* using the fixed-bottom line and hanging-long line methods in Panagatan Cays, Caluya, Antique, Philippines. *Journal of Applied Phycology, 8(2),* 105–109.

Hurtado-Ponce, A.Q., Agbayani, R.F., Sanares, R., & Castro-Mallare, M.T.R. (2001). The seasonality and economic feasibility of cultivating *Kappaphycus alvarezii* in Panagatan Cays, Caluya, Antique Philippines. *Aquaculture, 199,* 295–310.

Johnson, B., & Gopakumar, G. (2011). Farming of the seaweed *Kappaphycus alvarezii* in Tamil Nadu coast - status and constraints. *Marine Fisheries Information Service Technical and Extension Series, 208,* 1–5.

Khambhaty, Y., Mody, K., Gandhi, M.R., Thampy, S., Maiti, P., Brahmbhatt, H., Eswaran, K., & Ghosh PK (2012). *Kappaphycus alvarezii* as a source of bioethanol. *Bioresource Technology, 103(1),* 180–185.

Kotiya, A.S., Gunalan, B., Parmar, H.V., Jaikumar, M., Dave Tushar, D., Solanki, J.B., & Makwana, N.P. (2011) Growth comparison of the seaweed *Kappaphycus alvarezii* in nine different coastal areas of Gujarat coast, India. *Advances in Applied Science Research, 2(3),* 99–106.

Krishnamurty, V. Seaweeds - Wonder plants of the sea. *Aquaculture Foundation of India*, Chennai. 2005; 29 p.

Krishnan, M., & Kumar, R.N. (2010). Socio-economic dimensions of Seaweed Farming in India. Central Marine Fisheries Research Institute Special Publication 104: 1–78.

Luxton, L.M., Robertson, M., & Kindley, M.J. (1987). Farming of *Eucheuma* in the south Pacific islands of Fiji. *Hydrobiologia, 151/152,* 359–362.

Mairh, O.P., Soe-Htun, U., & Ohno, M. (1986). Culture of *Eucheuma striatum* (Rhodophyta, Solieriaceae) in subtropical waters of Shikoku, Japan. *Botanica Marina, 29*, 185–191.

Mairh, O.P., Zodape, S.T., Tewari, A., & Rajyaguru, M.R. (1995). Culture of Marine Alga *Kappaphycus striatum* (Schmitz) Doty on the Saurashtra region, Westcoast of India. *Indian Journal of Marine Sciences, 24*, 24–31.

Meinita, M.D.N., Kang, J.Y., Jeong, G.T., Koo, H.M., Park, S.M., & Hong, Y.K. (2012). Bioethanol production from the acid hydrolysate of the carrageenophyte *Kappaphycus alvarezii* (cottonii) *Journal of Applied Phycology, 24(4)*, 857–862.

Mishra, P.C., Jayasankar, R., & Seema, C. (2006). Yield and quality of carrageenan from *Kappaphycus alvarezii* subjected to different physical and chemical treatments. *Seaweed Research and Utilization, 28(1)*, 113–117.

Mollion, J., & Braud, J.P. (1993). A *Eucheuma* (Solieriaceae, Rhodophyta) cultivation test on the south-west coast of Madagascar. *Hydrobiologia, 260/261*, 373–378

Msuya, F.E. (2013). Effects of stocking density and additional nutrients on growth of the commercially farmed seaweeds Eucheuma denticulatum and *Kappaphycus alvarezii* in Zanzibar Tanzania. *Tan J Nat Appl Sci, 4(1)*, 605–612.

Msuya, F.E., & Kyewalyanga, M.S. (2006). Quality and quantity of phycocolloid carrageenan in the seaweeds *Kappaphycus alvarezii* and *Eucheuma denticulatum* as affected by grow out period, seasonality, and nutrient concentration in Zanzibar, Tanzania. Report submitted to Cargill Texturizing Solutions, 46pp

Msuya, F.E., Buriyo, A., Omar, I., Pascal, B., Narrain, K., Ravina, J.J.M., Mrabu, E., & Wakibia, J.G. (2014). Cultivation and utilisation of red seaweeds in the Western Indian Ocean (WIO) Region. *Journal of Applied Phycology, 26*, 699–705.

Muñoz, J., Freile-Pelegrín, Y., & Robledo, D. (2004). Mariculture of *Kappaphycus alvarezii* (Rhodophyta, Solieriaceae) color strains in tropical waters of Yucatán, México. *Aquaculture, 239*, 161–177.

National Academy of Agriculture Sciences (NAAS) India, Policy paper no. 22, NAAS, 2003; http://naasindia.org/Policy%20Papers/pp22.pdf [accessed 21.06.13]

Neish, I.C. (2003). The ABC of Eucheuma Seaplant Production-Agronomy, Biology and Crop-handling of *Betaphycus, Eucheuma* and *Kappaphycus* the *Gelatinae, Spinosum* and *Cottonii* of Commerce. SuriaLink Infomedia. Seaplants. Available from: http://www.fishdept.sabah.gov.my/download/ABC_eucheuma_a.pdf. [accessed 19.11.13.].

Neish, I.C. (2008). Good argonomy practices for *Kappaphycus* and *Eucheuma*: Including an overview of basic biology. SEAPlant.net Monograph HB2F 1008 V3 GAP, 72 pp

Nursidi, Ali, S.S., Niartiningsih, A., & Anshary, H. (2013). Relationship between environmental parameters and seaweed growth of *Kappaphycus alvarezii* cultivated in Saugi island pangkep regency South Sulawesi, e-journal of the Graduate School of Hasanuddin University, Indonesia, Published by Graduate School Hasanuddin University; Available from: http://pasca.unhas.ac.id/jurnal/files/ce4aa9844fd807a85578fd8e354d8960.pdf [accessed 24.12.13]

43

Ohno, M., Largo, D.B., & Ikurnoto, T. (1994). Growth rate, carrageenan yield and gel properties of culture kappa carrageenan producing red alga *Kappaphycus alvarezii* (Doty) Doty in the subtropical waters of Shikoku, Japan. *Journal of Applied Phycology, 6,* 1–5.

Ohno, M., Nang, H.O., & Hirase, S. (1996). Cultivation and carrageenan yield and quality of *Kappaphycus alvarezii* in the waters of Vietnam. *Journal of Applied Phycology, 8,* 431–437.

Ohno, M., Nang, H.O., Dinh, N.H., & Triet, V.D. (1995). On the growth of cultivated *Kappaphycus alvarezii* in Vietnam. *Japanese Journal of Phycology (Sôrui), 3,* 19–22.

Paula, E.J., & Pereira, R.T.L. (2003). Factors affecting growth rates of *Kappaphycus alvarezii* (Doty) Doty ex P. Silva (Rhodophyta, Solieraceae) in subtropical waters of São Paulo. *Brazilian Proceeding of International Seaweed Symposium, 17,* 381–388.

Periyasamy C, Anantharaman P, Balasubramanian T (2014a) Social upliftment of coastal fisher women through seaweed (*Kappaphycus alvarezii* (Doty) Doty) farming in Tamil Nadu, India. *Journal of Applied Phycology, 26(2)*, 775–781

Periyasamy C, Anantharaman P, Subba Rao PV (2014b) Experimental farming of *Kappaphycus alvarezii* (Doty) Doty with income estimates at different sites in the Mandapam region, Palk Bay, southeast coast of India. *Journal of Applied Phycology,* DOI 10.1007/s10811-014-0380-9.

Qian, P.Y., Wu, C.Y., Wu, M., & Xie, Y.K. (1996). Integrated cultivation of the red alga *Kappaphycus alvarezii* and the pearl oyster *Pinctada martensi. Aquaculture, 147,* 21–35.

Reis, R.P., Bastos, M., & Góes H.G. (2007). Cultivo de *Kappaphycus alvarezii* no litoral do Rio de Janeiro. *Panorama da AQÜICULTURA, janeiro/fevereiro, 17(89),* 42–47.

Santelices, B. (1999). A conceptual framework for marine agronomy. *Hydrobiologia 398/399,* 15–23

Sivasubramanian, V. (2011). Landscape Presentation on Algae and algal biotechnology for bioenergy-Indian contribution. BBSRC-DBT Bioenergy Workshop conducted on 10–11th October 2011 at International Centre for Genetic Engineering and Biotechnology(Delhi, India); http://www.rcuk.ac.uk/documents/india/Algae-India.pdf [accessed 21.06.13.]

Strickland, J.D.H., & Parsons, T.R. (1972). A practical handbook of seawater analysis. *Bulletin - Fisheries Research Board of Canada, 1,* 167–311.

Subba Rao, P.V., Kumar, K.S., Ganesan, K., & Thakur, M.C. (2008). Feasibility of cultivation of *Kappaphycus alvarezii* (Doty) Doty at different localities on the Northwest coast of India. *Aquaculture Research, 39,* 1107–1114.

Thirumaran, G., & Anantharaman, P. (2009). Daily Growth Rate of Field Farming Seaweed *Kappaphycus alvarezii* (Doty) Doty ex P. Silva in Vellar Estuary. *World Journal of Fish and Marine Sciences, 1(3),* 144–153.

Trono, G.C., & Ohno, M. (1989). Seasonality in the biomass production of the *Eucheuma* strains in Northern Bohol, Philippines. In: Umezaki I, editor. Scientific Survey of Marine Algae and their Resources in the Philippine Islands. Monbushio International Scientific Research Program, Japan. 71–80.

Villanueva, R.D., Romero, J.B., Montaño, N.E.M., & de la Peña, P.O. (2011). Harvest optimization of four *Kappaphycus species* from the Philippines. *Biomass bioenergy, 35,* 1311–1316.

Chapter II
Nutritional composition of Kappaphycus alvarezii

1. Introduction

Benthic marine macroalgae, commonly known as seaweeds, are one of the living renewable resources of the marine environment with potential food and therapeutic applications; they have been used directly or indirectly as human food in Asian countries and are considered under-exploited resources (Tseng 2004). Currently there is increasing consumer interest in products that can support or even promote health. Nutritionists acclaim seaweed as being low in calories, and rich in vitamins, minerals, and dietary fibers (Jimenez-Escrig and Sanchez-Muniz 2000).

Seaweeds have been conventionally used for a wide variety of application such as food, fodder, fertilizer, for medicinal purpose, and phycocolloids; moreover, few of them are also consumed as food in several Asian countries. Approximately, 250 species of seaweeds have been commercially utilized worldwide, amongst which 150 species are favorably consumed as human food; however, in western countries they form a source of polysaccharides (agar, alginates, carrageenans) for food and pharmaceutical industry (Zemke-White and Ohno 1999; Kumari et al., 2010). In order to cope with the growing market demand of seaweeds, several seaweeds are cultivated commercially as a source of livelihood worldwide.

Seaweeds generally show great variation in the nutrient contents, which could be related to several environmental factors such as water temperature, salinity, light and nutrients. Further, most of the environmental parameters influencing seaweed composition generally vary with season; moreover, the changes in ecological conditions can also stimulate or inhibit the biosynthesis of several nutrients (Soriano et al., 2006). Seasonal variations in the chemical composition and nutritive value have been reported in common marine seaweed from Hong Kong (Kaehler and Kennish 1996) and Ireland (Mercer et al., 1993). Though there have been several reports on biochemical composition of various seaweeds across India, but there are very few studies focusing on the temporal variations in chemical composition of seaweeds in Indian context. This study aims at presenting the seasonal variation in the proximate and mineral composition of tropical Indian red seaweed *K. alvarezii*, a carrageenophyte with enormous commercial value.

2. Materials and Methods

2.1. Seaweed: sampling and processing

Kappaphycus alvarezii was collected once in a month (from September 2004 to April 2006) from the cultivation site at Okha (22°28.656′ N and 69°04.015 ′E) Gujarat, Northwest coast of India. The samples were thoroughly washed with seawater followed by fresh water, and subsequently dried at 60°C in an oven, the dried samples were ground to particle size < 1mm and stored at room temperature in airtight plastic containers for further use.

2.2. Analysis of proximate composition

The nitrogen content of the dried seaweed was quantified using Kjeldahl procedure (Wathelet 1999) using KEL PLUS-KES 20L Digestion unit attached to a KEL PLUS–CLASSIC DX Distillation unit (M/s PELICAN Equipments, Chennai, India). The digestion was performed using sulfuric acid (96% H_2SO_4; initially for 75 min at 420°C, and again for 75 min at 370°C); this was thereafter distilled with boric acid solution (2%) and titrated with 0.1 M HCl. The crude protein content was estimated by multiplying the nitrogen by a factor of 6.25; while the total carbohydrate content was assayed by the phenol–sulphuric acid method using glucose as a standard (Dubois et al., 1956). Crude lipids were extracted using soxhlet extractor using 2:1 ratio of chloroform-methanol (Bligh and Dyer 1959). The fiber content of the dried seaweed was determined using a standard method outlined by AOAC (1990), i.e. the acid hydrolysis was carried out with sulfuric acid (0.3 N H_2SO_4) and the base hydrolysis was undertaken using sodium hydroxide (0.5 N NaOH). The cold extraction was performed with acetone; the sample was then dried (1 h at 110°C) until it reached a constant weight, cooled in a desiccator, and weighed (W_1); thereafter, it was placed in a muffle furnace at 550°C for 3 h, cooled (in a desiccator) and reweighed (W_2). The crude fiber percentage was calculated following the equation: % crude fiber = (W_1 – W_2 /W_0) × 100 (wherein, W_0 was the initial weight of the dried seaweed which was 2 g). The ash content was analyzed by shade drying the samples at room temperature and later in an oven at 80 °C for 1 h, thereafter, one gram of the powdered sample was accurately taken in a crucible, ashed at 550 °C in muffle furnace for 6 h to a constant weight, and the ash obtained was then quantified gravimetrically (AOAC 1995).

2.3. Analysis of minerals by ICP- OES

One gram seaweed was ashed, moistened with 10 drops of distilled water (Milli-Q) and carefully dissolved in 3 ml HNO_3 (1:1 v/v), followed by heating at 100 – 120 °C till the solution totally evaporated. The crucible was returned to muffle furnace and ashed again for 1 h at 550 °C and cooled. Subsequently the ash was dissolved in 3ml of 10 M HCl (1:1 v/v), and the solution was filtered through Millipore syringe filter (0.25µm) into 50 ml volumetric flask and 2 ml 0.1N HCl was added to the filtrate and the final volume was made up to 50 ml using distilled water (Milli-Q) (Kumar et al., 2007). Determination of mineral contents (Na, K, Ca, Mg, P, S, B, Cd, Co, Cr, Cu, Fe, Mn, Zn, Hg, Mo and V) of each seaweed sample was carried out using Inductively Coupled Plasma Optical Emission Spectroscopy, ICP-OES (Perkin-Elmer, Optima 2000). The analysis of all the above minerals was carried out in triplicate. Mean and standard deviation were calculated. All the chemicals and solvents used for experiments were of analytical grade.

2.4. Environmental parameters

Selected environmental parameters were determined at the farming site during the period of experimental cultivation (September 2004 to April 2006). All determinations, including water sample collections, were conducted at a depth of 30 cm i.e. the same depth as the cultured seaweeds. Seawater temperature was recorded daily throughout the study period and the mean monthly temperature was obtained. Salinity was determined fortnightly and the average value was recorded. The seawater samples were subjected to fortnightly nutrient analysis; here, dissolved inorganic phosphate (P–PO$_4$) and dissolved inorganic nitrate (N–NO$_3$) were determined according to the method described by Strickland and Parsons (1972) followed by which average monthly values were calculated.

2.5. Extraction of semi-refined carrageenan

The pre–cleaned dried seaweed (10 g) was rinsed with fresh water at ambient temperature and treated with 200 ml of 8% KOH (cooking solution), in a container (glass) maintained at approximately 70°C for 3 h. After 3 h of processing had elapsed, the seaweed was removed from the hot aqueous KOH solution and drained (this KOH mixture containing the seaweed was handled with extreme caution as it was corrosive having a pH ranging from 12-14). The seaweed was then subjected to a series of fresh water washes to reduce the pH, and, to wash residual KOH from the seaweed. This process helps eliminate certain salts and residual saponification products created during the KOH cook/reaction. The processed seaweed (containing carrageenan) obtained after the final rinse, was then chopped, dried and ground (Rideout and Bernabe 1997); this processed seaweed powder referred to as semi-refined carrageenan was then used for further determinations.

The carrageenan yield (%) was determined according to the formula:

$$\text{Yield \%} = (W_c / W_m) \times 100$$

Where, Wc is the extracted carrageenan weight (g) and Wm is the dry powder or algal weight (g) used for extraction. The data are presented as the mean yield obtained from three replicates.

2.6. Gel strength measurement

In order to evaluate the gel strength of the product, 1 g of carrageenan was soaked in a solution of 1% KCl for 2 hours. This mixture was boiled (100°C) for 20 minutes, cooled and left undisturbed at room temperature to facilitate the formation of gel; thereafter, the gels were refrigerated at 10°C overnight (approx. 10 h) in a refrigerator. Gel strength was measured at 25°C using a Nikkansui type gel tester (Kiya Seisakusho Ltd. Tokyo, Japan) as described by Hurtado-Ponce and Umezaki (1988) and it was expressed as g•cm^{-2}.

2.7. Statistical analysis

Analysis of variance (ANOVA) was conducted to determine the differences in seasonal variations of biochemical constituents; moreover multiple comparison tests with the least significance difference (LSD) were performed to determine significant differences.

3. Results and discussion

Seasonal fluctuations in various environmental parameters were observed during the study period. Apart from showing seasonal fluctuations in environmental parameters along the various coasts of India, several chemical oceanographic studies also reveal seasonal variations in nutrients and their interrelationships in the Northeastern Arabian Sea, and particularly the West coast of India (Sen Gupta et al., 1976). However, there are few studies highlighting the seasonal fluctuation in the various parameters influencing seaweed composition particularly the shoreline water of Okha coast. During the course of study a mean salinity and temperature were 34.20 ‰ and 26.55 °C respectively, which corresponded with reports of Gunalan et al., (2010) and Subba Rao et al., (2008). The seawater temperature varied from 22.5 to 30.1 °C, while the salinity fluctuated from 31.20 to 36.64‰. The tropical climate, warm, nutrient-enriched seawater, high light levels and high degree of water motion waters are suitable for farming *Kappaphycus* species, therefore, it could be projected that the prevalent environmental conditions encountered during this study were suitable for the growth of *Kappaphycus alvarezii*.

As evidenced in most studies, environmental parameters and season had a definite impact on the growth and biochemical constituent of *K. alvarezii*; Fig. 1 and 2 elucidate the seasonal variation in protein and carbohydrate content of *K. alvarezii* ($p < 0.001$) respectively.

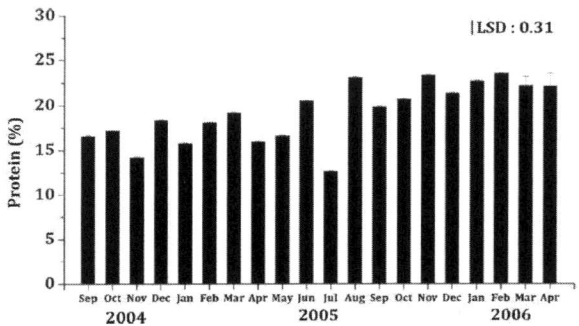

Fig. 1 Seasonal variation in protein content (%) of *K.alvarezii*

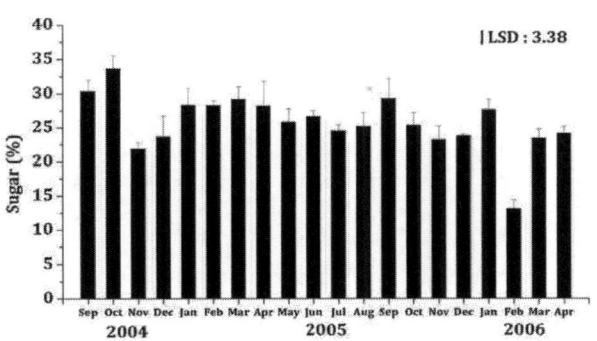

Fig. 2 Seasonal variation in sugar content (%) of *K.alvarezii*

Subba Rao et al., (2008) reported that the growth rate of *Kappaphycus alvarezii* significantly correlates with salinity, nitrate and seawater temperature at various sites; however, this study reveals that various environmental parameters had a vital impact on the chemical composition of this seaweed. The protein content of *K. alvarezii* ranged from 12.69 ± 0.6 to 23.61 ± 0.02 g/100g DW, these extensive seasonal fluctuations in the protein content of *K. alvarezii* could probably be attributed to the fluctuating environmental conditions. The average protein content of this seaweed (19.25 ± 0.15 g/100g DW) was much higher than that reported for *Sargassum vulgare* (4.59 – 9.97 g/100g DW; (Soriano et al., 2006), *Ulva lactuca* (7.06 ± 0.06 g/100g DW; (Wong and Cheung 2000) , *Undaria pinnatifida* (7.5 ± 1.9 g/100g DW; (Dawczynski et al., 2007), *Caulerpa veravelensis* (7.77 ± 0.59 g/100g DW; (Kumar et al., 2011), *C. lentillifera* (9.26 ± 0.03 g/100g DW; (Nguyen et al., 2011), *Caulerpa scalpelliformis* (10.50 ± 0.91 g/100g DW; (Kumar et al., 2011), *Laminaria* sp. (11.6 ± 0.8 g/100g DW; (Dawczynski et al., 2007), *C. racemosa* (12.88 ± 1.17 g/100g DW; (Kumar et al., 2011), *Hizikia fusiforme* (13.8 ± 6.2 g/100g DW; (Dawczynski et al., 2007) and *Ulva rigida* (17.8 g/100g DW; (Taboada et al., 2010). However, these values were comparable with that of *Hypnea charoides* (18.4 ± 0.30 g/100g DW; (Wong and Cheung 2000), *Hypnea japonica* (19.0 ± 0.3 g/100g DW; (Wong and Cheung 2000), and *Gracilaria cervicornis* (14.3 – 22.71 g/100g DW; (Soriano et al., 2006). El Din and El-Sherif (2012) reported the protein content of *C. prolifera*, *C. racemosa*, *C. bursa*, *H. tuna*, *U. petiolata*, *Udotea* sp., *G. verrucosa*, *R. ardissonei*, *C. spinosa*, *D. dichotoma*, *S. acinarium*, *S. hornschuchii* and *S. vulgare* to be 3.86, 8.23, 5.35, 11.8, 11.5, 14.5, 7.08, 4.5, 7.01, 8.15, 3.9, 6.05 and 6.15 g/100g DW respectively, however, these values are definitely lower than that of *K. alvarezii* (present study). Moreover, *K. alvarezii* contained relatively higher protein as compared to the seaweeds species habitually consumed as food (Fleurence 1999). Dawes (1998) vividly state that the members of Rhodophyta are characterized by greater protein content when compared to those of Phaeophyta, this could probably explain the high protein content in the species studied herein. Additionally, the mean of protein content of *K. alvarezii* (19.25 ± 0.15 g/100g DW) was higher

49

than that reported for higher plants (Norziah and Ching 2000). However, reports suggest that although chemical composition of seaweeds could be influenced by factors such as species, habitat, maturity, geographical locations, environmental parameters, and season (Ito and Hori 1989, Fleurence 1999), it is essential to note the very presence of proteins and other vital nutrients such as high concentrations of essential amino acids (EAA) (e.g. argenine, aspartine and glutamine found in many seaweed species; Fleurence 1999) make them a promising candidate to be used in various food formulations or as a supplement. More essentially, even though there may be seasonal variations in the protein content and in turn the essential amino acids, however, these EAA are available throughout the year despite seasonal variations in their concentrations (Galland –Irmouli et al., 1999).

The average protein content of *K. alvarezii* was 19.25 ± 0.15 g/100g DW, was much less than the mean carbohydrate content was 25.87 ± 1.64 g/100g DW. The carbohydrate content of *K. alvarezii* (average 25.87 ± 1.64 g/100g DW) was higher than several seaweeds, for e.g. *Ulva rigida*, *Ulva lactuca*, *Caulerpa racemosa*, *Sargassum filipendula* and *Hypnea japonica* (Rosenberg and Ramus 1982), *C. prolifera*, *C. racemosa*, *C. bursa*, *H. tuna*, *U. petiolata*, *Udotea* sp., *G. corneum*, *G. verrucosa*, *R. ardissonei*, *C. spinosa*, *D. dichotoma*, *S. acinarium*, *S. hornschuchii* and *S. vulgare* (El Din and El-Sherif 2012) which are reported to contain 6.40, 7.06, 3.98, 3.73, 4.28, 9.00, 6.93, 7.43, 7.40, 8.60, 8.00, 12.03, 15.77, 5.00, 20.91, 8.60, 13.67, 16.50 and 14.30 g/100g DW of carbohydrate respectively. It is essential to note that, in the present study the occurrence of maximum carbohydrate content of *K. alvarezii* coincided with occurrence of maximum biomass, thereby suggesting a link between seaweed growth and carbohydrate content; similarly, Rosenberg and Ramus (1982) also related the carbohydrate synthesis to periods of maximum growth, increased photosynthetic activity and a reduction in protein content. Moreover, during the period of occurrence of maximum carbohydrate in *K. alvarezii* higher values of water temperature, salinity and sunlight intensity were recorded, which confirms the influence of these parameters on carbohydrate synthesis (Munda and Kremer 1977). However, a negative correlation was found in this study between total carbohydrate and phosphate ($r = -0.471$). On the other hand, an inverse relationship was observed between carbohydrate and protein content; Givernaud et al., (1993) observed a similar a pattern for several species of seaweeds.

50

The lipid content varied from 0.39 ± 0.04 to 0.91 ± 0.51 g/100g DW (Fig. 3) with an average of 0.64 ± 0.08 g/100g DW. The lipid content of marine macroalgae (seaweed) has been reported to generally range between 1 – 6 g/100g DW (Dawczynski et al., 2007). However, in the present investigation on the carragenophytic seaweed *K. alvarezii*, the lipid content ranged from 0.39 ± 0.04 to 0.91 ± 0.51 g/100g DW (Fig. 3) which was comparable with *Sargassum polyschides* (0.7 ± 0.09 g/100g DW), *Corallina officinalis* (0.3 ± 0.2 g/100g DW; Sara Marsham et al., 2005) as well as that of *Hizikia* sp. and Arame (0.7–0.9 g/100g DW; Kolb et al., 1999); however, it was lower than *Palmaria* sp. (1.8 ± 0.14 g/100g DW), *Hypnea elongata* (0.97 ± 0.07 g/100g DW) and *Undaria pinnatifida* (1.05 ± 0.01 g/100g DW) (Sanchez- Machado et al., 2004). Moreover, there was a positive correlation between the water temperature and the lipid content (r = 0.554).

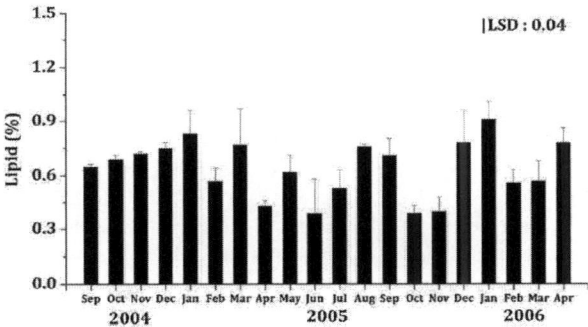

Fig. 3 Seasonal variation in lipid content (%) of *K.alvarezii*

Seaweeds are reported to be rich in fiber as compared to most fruits and vegetables; moreover, consumption of seaweeds helps increase the intake of dietary fiber and lower the occurrence of some chronic diseases (diabetes, obesity, heart diseases, cancers, etc.). In addition, consumption of seaweed is known to promote the growth of beneficial intestinal flora and also protect them; it also reduces the overall glycemic response, greatly increases stool volume and reduces the risk of colon cancer (Southgate 1990, Gupta and Abu-Ghannam 2011). *K. alvarezii* was particularly rich in fiber ranging from 9.68 ± 0.08 to 18.57 ± 0.15 g/100g DW (Fig. 4), which was higher than that reported by Robledo and Freile- Pelegrin (1997); they reported fiber content of *Sargassum vulgare* (mean 7.40 ± 1.61 g/100g DW) and *Gracilaria cervicornis* (mean 5.65 ± 0.74 g/100g DW) which were comparable with few other brown and red macroalgal species. On the other hand, the average fiber content of *K. alvarezii* (14.52 ± 0.11 g/100g DW) was comparable with *Gelidiella acerosa* (13.45 ± 1.07 g/100g DW) and lower than *Sargassum wightii* (17 ± 1.19 g/100g DW) (Syad et al., 2013).

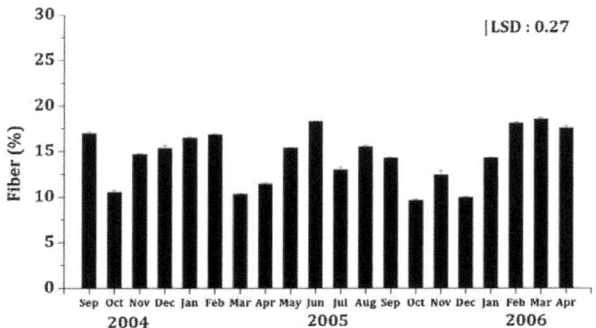

Fig. 4 Seasonal variation in fiber content (%) of *K.alvarezii*

The average SRC yield obtained in the present study (53.90 ± 1.37%) was comparable with western Pacific commercial sample (57%; Hayashi et al., 2007). Subba Rao et al., (2008), cite several reports mentioning lower semi-refined carrageenan content (31 – 43%) for the same seaweed farmed in the subtropical waters of Sao Paulo State, Brazil, and moderate yields for materials from Vietnam (34.5 to 45.3%; Ohno et al. 1996) and Indonesia (45%; Ohno et al., 1996), however, higher yields are reported from Philippines (54.6%; Trono and Ohno 1986). On the other hand, the SRC obtained in this study also demonstrated remarkable gel strength ranging from 348.33 ± 12.58 to 783.33 ± 15.28 g•cm^{-2} with an average of 602.00 ± 12.13 g•cm^{-2} which justifies its potential use in several pet food products. All the *K. alvarezii* samples yielded good amount of SRC i.e. the SRC yields ranged between 42.70 ± 1.07 to 63.73 ± 1.73 % (Fig. 5). A positive correlation could be derived between the carrageenan yield and gel strength (r = 0.578; Table 1); nonetheless, the nitrate present in seawater had a positive correlation with carrageenan yield (r = 0.465).

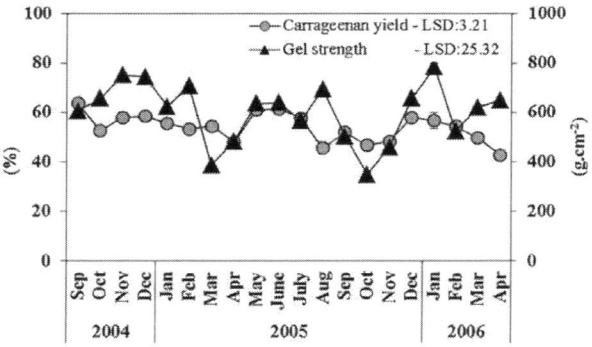

Fig. 5 Seasonal variation in semi-refined carrageenan yield (%) and gel strength (g·cm^{-2})

Table 1 Correlation coefficients (n=20) of the environmental parameters and the chemical composition of *K. alvarezii*

	Salinity	Phosphate	Nitrate	Biomass	Gel strength	Carrageenan Yield	Protein	Carbohydrate	Lipid	Fiber	Ash
Water °C	0.314	−0.169	0.139	0.005**	0.338	0.175	0.152	−0.042	0.554*	0.170	−0.259
Salinity		0.197	−0.216	0.193	−0.136	−0.426	0.373	−0.352	0.041	−0.127	0.077
Phosphate			0.380	−0.346	−0.278	−0.008	−0.174	−0.471*	−0.267	−0.076	0.071
Nitrate				−0.307	0.399	0.465*	−0.698**	0.004	0.055	0.017	−0.049
Biomass					−0.059	−0.196	0.140	−0.171	0.113	0.367	0.234
Gel strength						0.578*	−0.246	0.095	0.423	0.191	−0.155
Carrageenan yield							−0.476*	0.097	0.167	0.093	0.116
Protein								−0.355	0.023	0.144	0.046
Sugar									0.173	−0.313	−0.153
Lipid										0.052	0.196
Fiber											−0.205
Ash											1.000

*Significance (p<0.05)
**Significance (p<0.01)

The ash content of *K. alvarezii* ranged from 20.99 ± 0.08 to 33.81 ± 1.71 g/100g DW (Fig. 6). The ash contents of the seaweeds are usually much higher than land vegetables other than spinach (Rupeŕez et al., 2002; Omotoso 2006). However, the ash content is known to vary with species, geographical locations and season (Kaehler and Kennish 1996). The average ash content of *K. alvarezii* recorded in this study (27.00 ± 1.62 g/100g DW; fig. 6) was considerably higher than *Hypnea japonica* (22.1 ± 0.72 g/100g DW), *Hypnea charoides* (22.8 ± 2.23 g/100g DW), *Ulva lactuca* (21.3 ± 2.78 g/100g DW), *Hypnea pannosa* (15.3 g/100g DW), *Ulva lactuca* (24.6 g/100g DW), *Ulva pertusa* (24.7 g/100g DW), *Ulva lactuca* (11.0 ± 0.1 g/100g DW), *Durvillaea Antarctica* (17.9 ± 1.2 g/100g DW), *Caulerpa lentillifera* (22.20 ± 0.27 g/100g DW), *Gelidiella acerosa* (10.3 ± 4.9 g/100g DW) and *Sargassum wightii* (25 ± 2 g/100g DW) (Dawczynski et al., 2007, Mushollaeni 2011, Taboada et al., 2010, Syad et al., 2013).

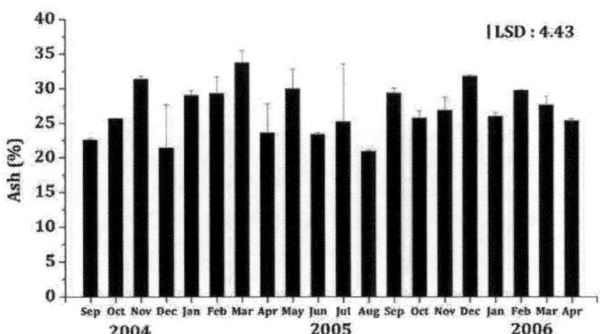

Fig. 6 Seasonal variation in ash content (%) of *K.alvarezii*

Generally, the ash content of a sample is known to reflect its mineral content. Mushollaeni (2011) state that the presence of ash content in seaweed showed that there were mineral salt in the sample; similarly, this study also confirms that *K. alvarezii* (which demonstrated high ash content) was also rich in minerals.

Seaweed are rich (8 – 40%) in several essential minerals and trace elements required for human nutrition (Indegaard and Ostgaard 1991); in fact, several reports suggest that the mineral content in seaweeds is higher than edible land plants (Ortega-Calvo et al., 1993; USDA 2001). However, it is obvious that occurrence and amounts of these may vary with species, moreover, within the same species there may also be seasonal fluctuations in the composition of these elements. Seasonal variation in mineral composition of *K. alvarezii* have been elucidated in Table 2; the highest mineral content (29939.61 ± 9340.38 mg/100g DW) was recorded in the month of April 2005, while the lowest value (10997.62 ± 1120.26 mg/100g DW) was recorded in January 2006. Among the 17 minerals analyzed, *K. alvarezii* demonstrated highest S content (11240 ± 730 mg/100g DW), and lowest quantities of Mo (0.04 ± 0.02 mg/100g DW). Apart from remarkable macro-mineral content (Na + K + Ca + Mg; 7910 ± 2950 mg/100g DW), *K. alvarezii* was also rich in selected micronutrients (Fe + Zn + Mn + Cu; 69.61 ± 0.16 mg/100g DW). The accumulation of Na and K salts generally depends upon the physiology as well as season; in this study the Na/K ratio of *K. alvarezii* ranged from 0.34 to 0.87. The macro-mineral content (Na + K + Ca + Mg) of *K. alvarezii* (7910 ± 2950 mg/100g DW) was lower than the values reported for the brown seaweeds such as *Fucus* (11723 ± 126 mg/100g DW), *Laminaria* (17061 ± 182 mg/100g DW), Wakame (17875 ± 382 mg/100g DW), and for the red seaweeds such as *Chondrus* (8606 ± 90 mg/100g DW) and Nori (8082 ± 214 mg/100g DW) (Indegaard and Ostgaard 1991); however, it was higher than many land vegetables (carrot 3276 mg/100g DW, sweet corn 1347 mg/100g DW, potato 6015 mg/100g DW, tomato 3429 mg/100g DW); however, spinach 9679 mg/100g DW is exceptional. Looking into the Mg content of all the samples, it could be stated that *K. alvarezii* could prove to be a good source for Mg, which is essential for regulating central nervous system excitability and normal functions. In this context, Syad et al., (2013) reported that low levels of Mg contribute to the heavy metal deposition in the brain that leads Parkinson's, multiple sclerosis and Alzheimer's disease; hence inclusion of this seaweed in the diet could evade these mentioned problems. Marine algae are also interesting candidates as Fe sources, especially in countries where the algal production is feasible; *K. alvarezii* also comprised significant quantities of Fe. In this study, the selected micronutrients (Fe + Zn + Mn + Cu) in the seaweed was higher (69.61 ± 0.16 mg/100g DW) than the land plant sweet corn (Fe + Zn + Mn + Cu, 4.9 mg/100g DW) and the seaweed *Laminaria* (5.1 mg/100g DW) (Food and Nutrition Board 1981).

K. alvarezii showed a sodium content relatively higher than the value reported for terrestrial vegetables (carrots, sweet corn, green peas, potato and tomato) (Food and Nutrition Board 1981). Generally, the intake of sodium chloride and diets with high Na/K ratio are related to the incidence of hypertension; here, the Na/K ratio of *K. alvarezii* ranged from 0.34 to 0.87. Ortega-Calvo et al., (1993) studied *Spirulina* and five other eukaryotic seaweeds used in food in

Spain, and reported Na/K ratios were below 1.5 in all seaweeds studied (0.33 – 1.34), which is interesting from the point of nutrition; nevertheless, the Na/K ratio in olives and sausages are 43.63 and 4.89, respectively (USDA 2001).

Seaweeds have been investigated for various elements and heavy metals in the past; in fact, reports suggest that most of the trace elements present in the algal biomass are heavy metals (As, Cd, Cu, Hg, Pb and Zn), but their content is generally below the toxic limit in most seaweeds (Ortega-Calvo et al., 1993). However, it is essential to quantify the mineral, ash and heavy metal content in any seaweed before recommending its use in food formulation. In the USA, algal products have to comply with the following maximum limits: 45% ash, 40 ppm heavy metals (Dietary Reference Intakes 2004). In the present study the Cu + Zn content was ranged from 1.47 to 4.31 mg/100g DW, which is below the toxic limits permitted for macro algae for human consumption (Ortega-Calvo et al., 1993). Indegaard and Ostgaard (1991) suggested incorporation of algal products in the daily diet of adults and recommend intake values of some trace elements (Fe: 10-18 mg, Zn: 15 mg, Mn: 2.5 –5 mg and Cu: 2 – 3 mg. 100 g dry weight).

Table 2 Macro (g/100 g DW), micro and trace elements (mg/100 g DW) of *K.alvarezii* determined by ICP-OES

Mineral	Sep-04	Oct-04	Nov-04	Dec-04	Jan-05	Feb-05	Mar-05	Apr-05	May-05	June-05
Na	2.49 ± 0.20	1.11 ± 0.29	2.56 ± 0.17	3.79 ± 0.17	3.27 ± 0.25	2.38 ± 0.36	2.61 ± 0.27	2.71 ± 0.31	2.05 ± 0.1	2.07 ± 0.17
K	3.40 ± 0.17	2.93 ± 0.21	4.08 ± 0.23	6.41 ± 0.17	5.27 ± 0.30	4.96 ± 0.45	5.85 ± 0.31	6.69 ± 0.26	3.46 ± 2.08	5.21 ± 0.2
Ca	0.82 ± 0.33	0.82 ± 0.17	0.80 ± 0.06	0.96 ± 0.12	0.80 ± 0.23	0.81 ± 0.15	0.83 ± 0.23	0.76 ± 0.42	0.98 ± 0.25	0.89 ± 0.06
Mg	0.74 ± 0.18	0.66 ± 0.24	0.69 ± 0.17	0.84 ± 0.12	0.91 ± 1.16	0.77 ± 0.71	0.71 ± 0.12	0.78 ± 0.15	0.92 ± 0.21	0.76 ± 0.12
P	0.10 ± 0.00	0.09 ± 0.00	0.11 ± 0.00	0.13 ± 0.00	0.17 ± 0.00	0.17 ± 0.00	0.12 ± 0.00	0.10 ± 0.00	0.13 ± 0.00	0.12 ± 0.00
S	5.43 ± 0.46	11.38 ± 0.20	11.37 ± 0.35	13.76 ± 0.52	13.06 ± 0.50	13.47 ± 0.79	15.39 ± 0.64	9.59 ± 0.46	12.13 ± 0.46	11.84 ± 0.46
Total	12.98 ± 1.13	16.96 ± 1.11	19.61 ± 0.98	25.89 ± 1.11	23.48 ± 2.44	22.56 ± 2.47	25.51 ± 1.57	20.63 ± 1.6	19.67 ± 3.1	20.89 ± 1.01
B	5.75 ± 0.02	1.74 ± 0.03	0.19 ± 0.01	0.60 ± 0.02	2.06 ± 0.02	5.68 ± 0.04	5.81 ± 0.02	6.14 ± 0.02	2.03 ± 0.02	4.56 ± 0.02
Cd	1.35 ± 0.01	1.54 ± 0.01	0.74 ± 0.02	0.80 ± 0.01	3.96 ± 0.03	2.07 ± 0.03	0.63 ± 0.02	0.83 ± 0.02	1.53 ± 0.01	1.85 ± 0.02
Co	0.48 ± 0.01	0.48 ± 0.01	0.71 ± 0.01	0.64 ± 0.02	0.20 ± 0.01	0.76 ± 0.03	0.19 ± 0.02	0.18 ± 0.02	2.65 ± 0.03	0.16 ± 0.03
Cr	2.66 ± 0.02	4.31 ± 0.02	5.36 ± 0.01	7.28 ± 0.01	0.74 ± 0.03	3.59 ± 0.02	1.73 ± 0.15	1.35 ± 0.04	13.47 ± 0.31	1.53 ± 0.03
Cu	0.62 ± 0.03	1.03 ± 0.02	1.07 ± 0.02	0.76 ± 0.01	0.76 ± 0.07	0.65 ± 0.03	0.58 ± 0.02	0.55 ± 0.06	3.77 ± 0.01	0.38 ± 0.01
Fe	49.16 ± 0.09	68.80 ± 0.06	94.50 ± 0.17	98.85 ± 0.01	132.73 ± 0.04	101.64 ± 0.08	34.57 ± 0.03	24.67 ± 0.02	111.83 ± 0.12	27.47 ± 0.25
Mn	1.35 ± 0.02	0.16 ± 0.02	2.26 ± 0.04	1.67 ± 0.01	0.66 ± 0.04	1.67 ± 0.02	1.03 ± 0.02	0.75 ± 0.03	1.24 ± 0.01	0.79 ± 0.01
Zn	2.21 ± 0.02	1.07 ± 0.02	2.26 ± 0.01	1.34 ± 0.03	1.65 ± 0.04	2.25 ± 0.02	2.17 ± 0.21	1.45 ± 0.02	2.26 ± 0.05	1.4 ± 0.02
Hg	0.02 ± 0.01	0.04 ± 0.01	0.22 ± 0.01	0.09 ± 0.01	0.04 ± 0.01	0.02 ± 0.01	0.02 ± 0.02	0.01 ± 0.01	0.08 ± 0.01	0.05 ± 0.01
Mo	0.05 ± 0.01	0.16 ± 0.01	0.01 ± 0.01	0.10 ± 0.01	0.05 ± 0.03	0.06 ± 0.05	0.03 ± 0.01	0.02 ± 0.12	0.03 ± 0.01	0.03 ± 0.01
V	0.11 ± 0.01	1.78 ± 0.02	0.33 ± 0.02	0.37 ± 0.12	0.04 ± 0.04	0.11 ± 0.09	0.11 ± 0.01	0.08 ± 0.01	0.1 ± 0.1	0.08 ± 0.01
Total	63.75 ± 0.24	81.11 ± 0.23	107.65 ± 0.32	112.49 ± 0.26	142.9 ± 0.35	118.51 ± 0.42	46.87 ± 0.50	36.03 ± 0.34	138.99 ± 0.68	38.3 ± 0.42
Grand Total (mg/100g)	13043.75 ± 1130.24	17041.11 ± 1110.23	19717.65 ± 980.32	26002.49 ± 1100.26	23622.90 ± 2440.35	22678.51 ± 2470.42	25556.87 ± 1570.50	20665.03 ± 1600.34	19808.99 ± 3100.68	20928.30 ± 1010.42

Table 2 (continued)

Mineral	July-05	Aug-05	Sep-05	Oct-05	Nov-05	Dec-05	Jan-06	Feb-06	Mar-06	Apr-06	Average	LSD
Na	2.06 ± 0.25	2.56 ± 1.53	2.13 ± 0.15	2.71 ± 0.15	2.34 ± 0.93	1.92 ± 0.61	1.39 ± 0.15	1.62 ± 0.15	1.13 ± 0.58	1.65 ± 0.38	2.23 ± 0.36	1.10
K	3.5 ± 0.12	7.59 ± 1.53	3.26 ± 0.38	3.28 ± 0.31	2.68 ± 0.25	2.78 ± 0.64	2.06 ± 0.1	3.01 ± 0.06	3.01 ± 0.15	2.42 ± 0.82	4.10 ± 0.44	2.60
Ca	0.89 ± 0.16	0.75 ± 2.52	0.97 ± 0.36	1.12 ± 0.35	0.92 ± 0.2	0.99 ± 0.15	0.76 ± 0.21	0.55 ± 0.17	0.53 ± 0.21	0.88 ± 0.18	0.84 ± 0.33	0.25
Mg	0.89 ± 0.46	0.72 ± 0.3	0.76 ± 0.16	0.77 ± 0.2	0.75 ± 0.21	0.76 ± 0.3	0.56 ± 0.2	0.65 ± 0.58	0.54 ± 0.15	0.66 ± 0.62	0.74 ± 1.82	0.25
P	0.17 ± 0.00	0.15 ± 0.00	0.12 ± 0.00	0.12 ± 0.00	0.25 ± 0.00	0.05 ± 0.00	0.09 ± 0.00	0.06 ± 0.00	0.10 ± 0.00	0.11 ± 0.00	0.12 ± 0.00	0.07
S	12.18 ± 0.17	18.03 ± 3.46	13.67 ± 2.04	11.3 ± 0.79	9.24 ± 0.35	10.36 ± 0.76	6.12 ± 0.46	9.23 ± 0.3	8.96 ± 0.79	8.24 ± 0.61	11.24 ± 0.73	5.23
Total	19.65 ± 1.16	29.8 ± 9.34	20.91 ± 3.09	19.3 ± 1.8	16.18 ± 1.94	16.86 ± 2.46	10.98 ± 1.12	15.12 ± 1.26	14.27 ± 1.88	13.96 ± 2.61	19.27 ± 3.68	
B	5.65 ± 0.02	4.67 ± 0.02	4.87 ± 0.02	2.16 ± 0.04	0.16 ± 0.03	0.93 ± 0.04	0.33 ± 0.06	1.07 ± 0.12	1.23 ± 0.06	0.63 ± 0.05	2.81 ± 0.03	3.70
Cd	0.92 ± 0.02	0.58 ± 0.02	1.03 ± 0.12	0.67 ± 0.03	0.4 ± 0.01	0.22 ± 0.02	0.23 ± 0.06	0.24 ± 0.01	0.23 ± 0.02	0.23 ± 0.02	1.06 ± 0.03	1.47
Co	2.14 ± 0.02	0.28 ± 0.02	0.6 ± 0.02	0.47 ± 0.06	0.76 ± 0.02	0.13 ± 0.06	0.1 ± 0.03	0.14 ± 0.01	0.06 ± 0.01	0.12 ± 0.03	0.56 ± 0.02	1.11
Cr	13.54 ± 0.01	5.34 ± 0.04	5.26 ± 0.02	1.57 ± 0.02	5.16 ± 0.03	0.87 ± 0.03	1.14 ± 0.0	1.33 ± 0.23	0.37 ± 0.01	1 ± 0.02	3.88 ± 0.05	6.3
Cu	0.65 ± 0.02	0.58 ± 0.02	0.55 ± 0.03	0.55 ± 0.03	0.77 ± 0.02	0.6 ± 0.1	0.35 ± 0.02	0.22 ± 0.01	0.19 ± 0.02	0.47 ± 0.02	0.76 ± 0.03	1.22
Fe	64.78 ± 0.01	123.6 ± 0.17	90.2 ± 0.02	45.5 ± 0.03	126.77 ± 0.23	48.56 ± 0.02	13.83 ± 0.03	21.15 ± 0.02	9.09 ± 0.02	31.19 ± 0.02	65.94 ± 0.07	81.78
Mn	0.69 ± 0.01	1.6 ± 0.02	1.28 ± 0.04	1.17 ± 0.04	1.48 ± 0.02	1.25 ± 0.02	0.5 ± 0.01	0.57 ± 0.01	0.36 ± 0.02	0.88 ± 0.02	1.10 ± 0.02	0.88
Zn	1.43 ± 0.02	2.76 ± 0.03	1.8 ± 0.1	1.77 ± 0.02	3.37 ± 0.02	2.96 ± 0.04	1.04 ± 0.01	1.09 ± 0.01	0.81 ± 0.02	1.99 ± 0.03	1.85 ± 0.04	1.16
Hg	0.07 ± 0.0	0.05 ± 0.0	0.01 ± 0.0	0.01 ± 0.01	0.1 ± 0.01	0.02 ± 0.01	0.02 ± 0.03	0.01 ± 0.01	0.02 ± 0.01	0.02 ± 0.01	0.05 ± 0.01	0.08
Mo	0.02 ± 0.01	0.01 ± 0.01	0.04 ± 0.02	0.07 ± 0.01	0.01 ± 0.01	0.05 ± 0.01	0.03 ± 0.0	0.03 ± 0.01	0.02 ± 0.01	0.04 ± 0.01	0.04 ± 0.02	0.06
V	0.08 ± 0.01	0.14 ± 0.03	0.16 ± 0.02	0.13 ± 0.01	0.15 ± 0.02	0.13 ± 0.01	0.05 ± 0.01	0.07 ± 0.01	0.05 ± 0.02	0.09 ± 0.01	0.21 ± 0.03	0.62
Total	89.97 ± 0.15	139.61 ± 0.38	105.8 ± 0.41	54.07 ± 0.3	139.13 ± 0.42	55.72 ± 0.36	17.62 ± 0.26	22.92 ± 0.45	12.43 ± 0.22	36.66 ± 0.24	78.20 ± 0.35	
Grand Total (mg/100g)	19739.97 ± 1160.15	29039.61 ± 9340.38	21015.80 ± 3090.04	16399.13 ± 1940.42	16915.72 ± 2460.36	10907.62 ± 2460.36	15145.92 ± 1120.26	14282.43 ± 1260.45	13996.66 ± 1880.22	± 2610.24		

Table 3 provides a comparison of the various recommended dietary allowances of macronutrients, minerals and trace elements and their presence in *K. alvarezii* (1 g); as elucidated this seaweed indeed qualifies to be incorporated as a food ingredient.

Table 3 Recommended dietary allowances of macronutrients, minerals and trace elements of *K. alvarezii*

Macronutrients, minerals and trace elements	Recommended Dietary Allowances per day[a]	Average content per gram in *K. alvarezii*
Macronutrients (g)		
Protein	9–71	0.17
Carbohydrate	60–210	0.23
Fiber	19–29	0.12
Fats	ND	0.007
Macro elements (mg)		
Boron	<20	0.03
Calcium	210–1,000	8.42
Iron	6–27	0.63
Magnesium	30–320	7.42
Manganese	0.003–2.6	0.011
Phosphorus	100–1,250	1.23
Zinc	2–12	0.019
Microelements (µg)		
Cadmium	50–150	10.02
Chromium	5.5–45	38.87
Copper	200–1,300	7.55
Molybdenum	2–50	43
Mercury	40	0.46

[a] According to Dietary Reference Intakes (2004), Food and Nutrition Board (1989), Recommended Daily Intakes of Various Food Supplements (2007), Concon (1988) and Phaneuf et al. (1999)

Based on the study conducted herein, it could be stated that *K. alvarezii* could be an excellent source of essential macronutrients and minerals, therefore, it could probably compensate for the frequently low minerals content of food; moreover, it also comprises significant quantities of protein and fiber. Hence, it could be preferentially used as an essential value–added food supplement or spice to the vegetarian or omnivorous diet.

4. Conclusion

Seaweeds have been traditionally used for variegated purposes including several food applications. Studies on temporal variations in the proximate and mineral composition of *K.*

alvarezii conducted herein reveal the consistent presence of fiber and minerals (micro and macro nutrient) in this widely cultivated seaweed, thereby suggesting its use as a potential food supplement. *K. alvarezii* having high nutritional value, could either be used in the food industry as a source of these essential ingredients, or could be used as such for oral consumption after proper processing. However, its commercial value could also be enhanced by improving its quality and further increasing the probability of its use in a wider range of novel seaweed-based products. Nevertheless, it could definitely be incorporated into various animal and pet food products as a vital value additive supplement.

References

AOAC, (1990). Official Methods of Analysis, Association of Official Analytical Chemists, 15th edn. AOAC Press, Gaithersburg, USA.

AOAC, (1995). Official Methods of Analysis. In: Horwitz, W. (ed) Association of Official Analytical Chemists, Washington, DC, USA.

Bligh, E.G., & Dyer, W.J. (1959). A Rapid method of Total lipid Extraction and Purification. *Canadian Journal of Biochemistry and Physiology, 37,* 911–917.

Concon, J.M. (1988). Food Toxicology, Part A (Vol. 582). New York: Marcel Dekker Inc., pp. 1049–1073.

Dawczynski, C., Schubert, R., & Jahreis, G. (2007). Amino acids, fatty acids, and dietary fiber in edible seaweed products. *Food Chemistry, 103,* 891–899.

Dawes, C.J. (1998). Marine Botany. John Wiley & Sons, Inc., New York.

Dietary Reference Intakes, (2004). Food and Nutrition Board, Institute of Medicine, National Academies, Washington DC. Available from: http://www.iom.edu

Dubois, M., Gilles, K.A., Hamilton, J.K., Rebers, P.A., & Smith, F. (1956). Colorimetric methods for determination of sugars and related substances. *Analytical Chemistry, 28,* 350–356.

El Din, N.G.S., & El-Sherif, Z.M. (2012). Nutritional value of some algae from the north-western Mediterranean coast of Egypt. *Journal of Applied Phycology, 24,* 613–626.

Fleurence, J. (1999). Seaweed proteins: biochemical, nutritional aspects and potential uses. *Trends in Food Science & Technology, 10,* 25–28.

Food and Nutrition Board, (1981). Food chemical codex, 3rd ed. National Academy Press, Washington, DC.

Food and Nutrition Board, (1989). National Academy of Sciences-Recommended Dietary Allowances, Revised. Available from: http://www.diet-and-health.net/Nutrients/rdas.html

Galland-Irmouli, A.V., Fleurence, J., Lamghari, R., Lucon, M., Rouxel, C., Barbaroux, O., Bronowicki, J.P., Villaume, C., & Guéant, J.L. (1999). Nutritional value of proteins from edible seaweed *Palmaria palmata* (Dulse). *The Journal of Nutritional Biochemistry, 10,* 353–359.

Givernaud, M.A., Givernaud, T., Morvan, H., & Cosson, J. (1993). Annual variations of the biochemical composition of *Gelidium latifolium* (greville) Thuret et Bornet. *Hydrobiologia, 260/261,* 607–612.

Gunalan, B., Kotiya, A.S., & Jetani, K.L. (2010). Comparison of *Kappaphycus alvarezii* Growth at Two Different Places of Saurashtra Region. *European Journal of Applied Sciences, 2(1),* 10–12.

Gupta, S., & Abu-Ghannam, N. (2011). Bioactive potential and possible health effects of edible brown seaweeds. *Trends in Food Science & Technology, 22,* 315–326.

Hayashi, L., Paula, E.J.D. & Chow, F. (2007). Growth rate and carrageenan analyses in four strains of *Kappaphycus alvarezii* (Rhodophyta, Gigartinales) farmed in the subtropical waters of São Paulo State. *Brazilian Journal of Appl Phycology, 19,* 393–399.

Hurtado-Ponce, A.Q., & Umezaki, I. (1988). Physical properties of agar gel from *Gracilaria* (Rhodophyta) of the Philippines. *Botanica Marina, 31,* 171–174.

Indegaard, M., & Ostgaard, K. (1991). Polysaccharides for food and pharmaceutical uses. In: Guiry MD, Blunden G. (eds) Seaweed resources in Europe: uses and potential. John Wiley & Sons Ltd, Chichester, pp 169–183.

Ito, K., & Hori, K. (1989). Seaweed: chemical composition and potential uses. *Food Research International, 5,* 101–144.

Jimenez-Escrig, A., & Sanchez-Muniz, F.J. (2000). Dietary fiber from edible seaweeds: chemical structure, physicochemical properties and effects on cholesterol metabolism. *Nutrition Research, 20(4),* 585–598.

Kaehler, S, & Kennish, R. (1996). Summer and winter comparisons in the nutritional value of 447 marine macroalgae from Hong Kong. *Botanica Marina, 39,* 11–17.

Kolb, N., Vallorani, L., & Stocchi, V. (1999). Chemical composition and evaluation of protein quality by amino acid score method of edible brown marine algae Arame (*Eisenia bicyclis*) and Hijiki (*Hijikia fusiforme*). Acta Alimentaria, 28, 213–222.

Kumar, M., Gupta, V., Kumari, P., Reddy, C.R.K., & Jha, B. (2011). Assessment of nutrient composition and antioxidant potential of Caulerpaceae seaweeds. *Journal of Food Composition and Analysis, 24,* 270–278.

Kumar, S.K., Ganesan, K., & Subba Rao, P.V. (2007). Phycoremediation of heavy metals by the three-color forms of *Kappaphycus alvarezii*. *Journal of Hazardous Materials, 143,* 590–592.

Kumari, P., Kumar, M., Gupta, V., Reddy, C.R.K., & Jha, B. (2010) Tropical marine macroalgae as potential sources of nutritionally important PUFAs. *Food Chemistry, 120,* 749–757.

Mercer, J.P., Mai, K.S., & Donlon, J. (1993). Comparative studies on the nutrition of two species of abalone, *Haliotis tuberculata* Linnaeus and *Haliotis discus* hannaiIno. I. Effects of algal diets on growth and biochemical composition. *Invert Reprod Dev, 23,* 2–3.

Munda, I.M., & Kremer, B.P. (1977). Chemical composition and physiological properties of fucoids under conditions of reduced salinity. *Marine Biology, 42,* 9–15.

Mushollaeni, W. (2011). The physicochemical characteristics of sodium alginate from Indonesian brown seaweeds. *African Journal of Food Science, 5(6),* 349–352.

Nguyen, V.T., Ueng, J.P., & Tsai, G.J. (2011). Proximate composition, total phenolic content, and antioxidant activity of seagrape (*Caulerpa lentillifera*). *Journal of Food Science, 76,* C950–C958.

Norziah, M.H., & Ching, C.Y. (2000). Nutritional composition of edible seaweed *Gracilaria changgi*. *Food Chemistry, 68,* 69–76.

Ohno, M., Nang, H.O., & Hirase, S. (1996). Cultivation and carrageenan yield and quality of *Kappaphycus alvarezii* in the waters of Vietnam. *Journal of Applied Phycology, 8,* 431–437.

Omotoso, O.T. (2006). Nutritional quality, functional properties and anti-nutrient compositions of the larva of *Cirina forda* (Westwood) (Lepidoptera: Saturniidae). *Journal of Zhejiang University SCIENCE B, 7(1),* 51–55.

Ortega-Calvo, J.J., Mazuelos, C., Hermosin, B., & Saiz-Jimenez, C. (1993). Chemical composition of *Spirulina* and Eukaryotic algae food products marketed in Spain. *Journal of Applied Phycology, 5,* 425–435.

Phaneuf, D., Cote, I., Dumas, P., Ferron, L.A., & LeBlanc, A. (1999). Evaluation of the contamination of Marine Algae (Seaweed) from the St. Lawrence River and Likely to be consumed by Humans. *Environmental Research Section A, 80,* S175–S182.

Recommended daily intakes of various food supplements (2007) Lenntech Water & Luchtbehandeling Holding B.V., Netherlands. (Available from: http://www.lenntech.com/recommended-daily-intake.htm on 15th August 2007

Rideout, C.S., Bernabe, M.G., US patent 5,801,240, field March 7, 1997, and issued September 1, 1998.

Robledom, D., & Freile-Pelegrin, Y. (1994). Chemical and mineral composition of six potentially edible seaweed species of Yucatán. *Botanica Marina, 40,* 301–306.

Rosenberg, G., Ramus, J. (1982). Ecological growth strategies in the seaweeds *Gracilaria follifera* (Rhodophyceae) and *Ulva* sp. (Chlorophyceae): soluble nitrogen and reserve carbohydrates. *Marine Biology, 66,* 251–259.

Rupeŕez, P. (2002). Mineral content of edible marine seaweeds. *Food Chemistry, 79,* 23–26.

Sanchez-Machado, D.I., Lopez-Cervantes, J., Lopez–Hernandez, J., & Paseiro–Losada, P. (2004). Fatty acids, total lipids, protein and ash contents of processed edible seaweeds. *Food Chemistry, 85,* 439–444.

Sara Marsham, S., Scott, G.W., & Tobin, M.L. (2005). Comparison of nutritive chemistry of a range of temperate seaweeds. *Food Chemistry, 100,* 1331–1336.

Sen Gupta, R., Sankaranarayanan, V.N., De Sousa, S.N., & Fondekar, S.P. (1976). Chemical oceanography of the Arabian Sea. Part III: studies on nutrient fraction and stoichiometric relationships in the northern and eastern basins. *Indian Journal of Marine Sciences, 5,* 58–71.

Soriano, E.M., Fonseca, P.C., Carneiro, M.A.A., & Moreira, W.S.C. (2006). Seasonal variation in the chemical composition of two tropical seaweeds. *Bioresource Technology, 97,* 2402–2406.

Southgate, D.A.T. (1990). Dietary fiber and health. In: Southgate DAT, Waldron K, Johnson IT, Fen-wick GR (eds) Dietary fiber: Chemical and biological aspects. The Royal Society of Chemistry, Cambridge, pp 10–19.

Strickland, J.D.H., & Parsons, T.R. (1972). A Practical handbook of seawater analysis. *Bulletin - Fisheries Research Board of Canada, 67,* 1–311

Subba Rao, P.V., Kumar, K.S., Ganesan, K., & Thakur, M.C. (2008). Feasibility of cultivation of *Kappaphycus alvarezii* (Doty) Doty at different localities on the Northwest coast of India. *Aquaculture Research, 39,* 1107–1114.

Syad, A.N., Shunmugiah, K.P., & Kasi, P.D. (2013). Seaweeds as nutritional supplements: Analysis of nutritional profile, physicochemical properties and proximate composition of *G. acerosa* and *S. wightii. Biomedicine & Preventive Nutrition, 3,* 139–144.

Taboada, C., Millán, R., & Míguez, I. (2010). Composition, nutritional aspects and effect on serum parameters of marine algae *Ulva rigida. Journal of the Science of Food and Agriculture, 90,* 445–449.

Trono, G.C., & Ohno, M. (1986). Seasonality in the biomass production of the Eucheuma strains in Northern Bohol, Philippines. In: Umezaki I (ed) Scientific Survey of Marine Algae and their Resources in the Philippine Islands. Monbushio International Scientific Research Program, Japan, pp 71–80.

Tseng, C.K. (2004). The past, present and future ofphycology in China. *Hydrobiologia, 512,* 11–20.

USDA, (2001). Agricultural research service. Nutrient Database for Standard Reference, Release 14

Wathelet, B. (1999). Nutritional analysis for proteins and amino acids in beans (*Phaseolus* sp.). *Biotechnology, Agronomy, Society and Environment, 3,* 197–200.

Wong, K.H., & Cheung, P.C.K. (2000). Nutritional evaluation of some subtropical red and green seaweeds. Part I. Proximate composition, amino acid profiles and some physico-chemical properties. *Food Chemistry, 71,* 475–482.

Zemke-White, L.W., & Ohno, M. (1999). World seaweed utilisation: An end-of-century summary. *Journal of Applied Phycology, 11,* 369–376.

Chapter III
Functional Properties of Protein Concentrate of Kappaphycus alvarezii

1. Introduction

Proteins are responsible for many of the functional properties that influence the consumer acceptance of food products; therefore, they play a key role in food processing as well as in the development of food products (Ogunwolu, Henshaw, Mock, Santros & Awonorin, 2009). Based on the source, food proteins could be roughly grouped into animal proteins (e.g. gelatin, milk protein) and vegetable proteins (e.g. soya protein, peanut protein and wheat protein). The availability, cost and the risk factors associated with diseases from animal protein sources make the nutritionist consider alternative plant protein sources for human being and feedstock preparation. Vegetable proteins have been prevalently used in various food applications, owing to their acceptable functional properties, such as emulsification, fat and water absorption, texture modifications, color control and whipping properties. In the recent years, many plants have attracted a great deal of interest as a source of low-cost protein to supplement human diets. But although plant protein sources are generally cheaper as compared with the animal protein sources, these plant proteins supplements are lower in some essential amino acids, energy and minerals such as phosphorus as compared with animal protein supplements; at the same time they constitute certain anti-nutritional factors (Yun, Kwon, Lohakare, Choi, Yong, Zheng, Cho & Chae, 2005). The production of plant protein concentrates (PCs) is of growing interest to the food industry because of the increasing application of plant proteins in foods especially in the developing countries. Moreover, plants PCs are quite extensively used in food to improve the nutritional quality of the product for economic reasons, or as a functional ingredient (for e.g. the use of soyabean PCs and whey PCs (Wong, Cheung & PO Ang, 2004). The demand for relatively inexpensive sources of proteins that are incorporated into value-added food products is increasing globally. It is essential to consider that plant protein sources are susceptible to to climate change and require availability of agricultural land; therefore, people are in the quest for economically viable alternatives like biomass from the marine environment.

Seaweeds are used extensively as food in coastal cuisines around the world especially they have been traditionally used for human consumption in Asia, however, their use as animal fodder has been popular in Norway (Slaski & Franklin, 2011). Approximately, 250 species of seaweeds have been commercially utilized worldwide, amongst which 150 species are favorably consumed as human food; however, in western countries they form a source of polysaccharides (such as agar, alginates, or carrageenans) for food and pharmaceutical industry. From a nutritional point of view, edible seaweeds are low calorie food, having great nutritional value owing to their vitamin, protein and mineral contents; apart from constituting vitamin A, B_1, B_{12} and C; they are also natural sources of hydrosoluble and liposoluble vitamins, such as thiamine and riboflavin, β-carotene and tocopherols (Kumar & Kaladharan, 2007). Furthermore, they possess long-chain polyunsaturated essential fatty acids from the omega-3 family (e.g. eicosapentaenoic acid), thereby having potential use in the development of low-cost, highly nutritive diets for human and animal nutrition (Kumar & Kaladharan, 2007). In fact, due to their high essential amino acids content and relatively high level of unsaturated fatty acids, the quality of proteins and lipids in seaweeds are better suited for consumption as compared to other vegetables. Seaweeds are mostly used in human or animal foods keeping in mind their mineral contents or for the functional properties of their polysaccharides, and are rarely promoted for the nutritional value of their proteins (Fleurence, 1999).

Recently, people have been looking into the use of seaweeds as an economic alternative source for protein concentrates; additionally, reports suggest that the nutritional potential of seaweed as a source of protein has been known to vary with species (Fleurence, 1999). However, very few studies have been taken up on the quality of seaweed protein owing to the complications of extraction and preparation of protein concentrates (Wong, Cheung & PO Ang, 2004); in fact, barely any research has been performed on the protein of seaweeds for human consumption. To increase its utilization, there is a need to process the whole seaweed biomass into high protein products such as protein concentrate and isolates, and then examine the suitability of these products for use as functional ingredients and food supplements. However, the ultimate success of utilizing any plant protein as food ingredients depends largely on its functional and nutritional properties. Studies on the functional properties of proteins, such as solubility, water/oil holding capacity, emulsifying activity, foaming ability and stability, viscosity, and gelation, which are in turn highly dependent on many factors such as pH and type and amount of salt present etc., have been reported by several authors (Ganesan, Kumar & Subba Rao, 2012; Gbadamosi, Abiose & Aluko, 2012). Rhodophtyic seaweeds possess good amount of protein (Wong, Cheung & PO Ang, 2004); particularly *Kappaphycus alvarezii* (Doty) Doty constitutes significant amount of protein (Rajasulochana, Krishnamoorthy & Dhamotharan, 2012). Moreover, this edible seaweed having multifarious food applications, is also be used in pet food and as aquaculture feed. However, research on the functional properties of the protein concentrate of this seaweed has not been undertaken and brought to the public domain. The present study therefore aimed to investigate the functional properties of protein concentrate of this widely cultivated seaweed *Kappaphycus alvarezii* (Doty) Doty considering its use as an ingredient various in food formulations.

2. Materials and method

2.1. Sample preparation

Fresh *Kappaphycus alvaerezii* collected from cultivation farm, Port Okha (22°28.65′N and 69°04.01′E), Gujarat, Northwest coast of India, was sun-dried, and thoroughly washed with distilled water to remove epiphytes. This cleansed seaweed was then oven dried at 60 °C for 16 h to a constant weight. The dried moisture-free sample was then pulverized to obtain uniformly sized (0.5 mm) particles. The milled seaweed sample was then stored in airtight plastic bags in a desiccator at room temperature (25 °C) prior to extraction of the protein concentrate (PC).

2.2. Extraction of protein concentrate

K. alvaerezii PC was extracted according to a slightly modified methodology of Fleurence, Le Coeur, Mabeau, Maurice & Landrein (1995). In brief, seaweed powder was suspended in deionized water (1:20 w/v); this suspension was gently stirred overnight at 35°C. After incubation, the suspension was centrifuged at 10000 g at 4°C for 20 min. The supernatant was collected, and the pellet was re-suspended in de-ionized water in the presence of 0.5% (v/v) 2–mercaptoethanol. Then the pH of the mixture was adjusted to 12 with 1 M NaOH. The mixture was gently stirred at room temperature for 2 h before centrifugation under the same conditions stated above. The second supernatant was collected and combined with the previous supernatant. The combined supernatant was stirred at 0 ± 4°C, and its pH was adjusted to 7 before precipitation with solid ammonium sulphate. This extraction procedure (mentioned above) was repeated five times on the residue. The seaweed PCs were precipitated from the supernatant by gently adding solid ammonium sulphate along with stirring until 85% saturation (60 g/100 ml) was reached. Then this mixture was allowed to stand for 30 min before centrifugation under conditions mentioned above. The pellet (PC) obtained was dialyzed against distilled water until the total dissolved solutes (TDS) (mg/l) of dialysate, were similar to those of the distilled water. Finally, the retentate containing the seaweed PCs were freeze-dried, powdered, and stored in air-tight bags in desiccators before evaluation of its functional properties.

2.3. Determination of total protein content

The nitrogen content of the PC was determined by Kjeldahl method (Wathelet, 1999) using KEL PLUS–KES 20L Digestion unit attached to a KEL PLUS–CLASSIC DX Distillation unit (M/s PELICAN Equipment, Chennai, India); thereafter, the crude protein content of PCs was calculated by multiplying its nitrogen content by a factor of 6.25.

2.4. Nitrogen Solubility

Nitrogen solubility was determined by the method of Bera & Mukherjee (1989). Here, PC samples (100 mg each) were dispersed varying concentrations (0.1 and 0.5M) of NaCl solutions as well as in 5 ml of distilled water. The pH of the mixture was adjusted to 2 – 12 using 0.1 N HCl or NaOH. Samples were shaken at 145 rpm for 30 min at room temperature and then centrifuged at 4000 g for 30 min. Nitrogen contents of the supernatants were determined by Kjeldahl method, and percent nitrogen solubility was calculated as follows:

$$\text{Solubility (\%)} = \frac{\text{Amount of nitrogen in the supernatant}}{\text{Amount of nitrogen in protein concentrate}} \times 100$$

2.5. Water-holding capacity (WHC)

Water-holding capacity (g of H_2O/g of PC) was determined using the method of Bencini (1986). Protein concentrate (0.5 g) was transferred into a pre-weighed 15 ml centrifuge tube, and 10 ml of distilled water was added to it; this was then mixed at high speed using a vortex mixer (Tarson, India) for 2 min. After the mixture was uniformly wet and consistent, it was allowed to stand at room temperature for 30 min, and then centrifuged at 2000g for 30 min. The supernatant obtained thereby was decanted, and the centrifuge tube containing sediment was weighed. Water-holding capacity was calculated by the following formula.

$$WHC \ (g \ H_2O) = \frac{W_2 - W_1}{W_0} \ X \ 100$$

Where W_0 is the weight of the dry sample (g), W_1 the weight of the tube plus the dry sample (g), and W_2 weight of the tube plus the sediment (g).

2.6. Fat absorption capacities (FAC)

In order to determine the fat absorption capacities (g of oil/g of PC), 1.0 g of PC sample was taken in a pre-weighed centrifuge tube and thoroughly mixed with 5 ml of sunflower oil. This protein–oil mixture was then centrifuged (3000g for 30 min); immediately after centrifugation, the supernatant was carefully removed, and the tubes were weighed. FAC (grams of oil per gram of protein) was calculated as

$$FAC \ (g \ Oil) = \frac{F_2 - F_1}{F_0} \ X \ 100$$

Where F_0 is the weight of the dry sample (g), F_1 is the weight of the tube plus the dry sample (g), and F_2 is the weight of the tube plus the sediment (g).

2.7. Emulsifying and surface-active properties

Emulsifying activity was measured using a modified method of Cooper & Goldenberg (1987). Here, oil was added to aqueous phase containing the protein concentrate (10 mg/ml); here the hydrocarbon: PC ratio was 3:2 (v/v). This mixture was stirred vigorously for 2 min on a cyclo-mixer and thereafter left undisturbed. The oil, emulsion and aqueous layers were measured at different time intervals and an emulsification index (E) was calculated as follows

$$Emulsifying \ index \ (E) = \frac{Volume \ of \ the \ emulsion \ layer}{Total \ Volume \ of \ the \ mixture} \ X \ 100$$

The emulsification index was noted with respect to time (15, 30, 90, 210, 390, 720 min) and was represented accordingly, i.e. the emulsification index after 15, 30, 90, 210, 390 and 720 minutes was represented as E_{15}, E_{30}, E_{90}, E_{210}, E_{390} and E_{720} respectively.

In order to understand surface-active properties of the PC, the surface tension of 0.1 and 0.5% (w/v) the PC was determined using a Dataphysics Dynamic Contact Angle Meter and Tensiometer (DCAT 21), Dataphysics Instruments GmbH, Germany using Wilhelmy plate (PT 11) made of platinum-iridium.

2.8. Foaming capacity and stability

A modified method of Nath & Rao (1981) was used to determine the foaming capacity and stability of the protein concentrate. A 100 ml solution of the protein concentrate (20µg/ml) was using a vortex mixture for 5 min at room temperature and transferred to a measuring cylinder. The volume increase is expressed as percent foaming capacity.

$$\text{Foaming capacity (\%)} = \frac{V_2 - V_1}{V_1} \text{ X } 100$$

where, V_1 is the volume of protein solution before whipping, and V_2 is the volume of protein solution after whipping.

The foam stability was determined by measuring the decrease in volume of foam as a function of time up to a period of 90 min with an interval of 30 min at different pH level ranged from 2-10.

$$\text{Foam stability (\%)} = \frac{\text{Volume after standing - Volume before whipping}}{\text{Volume before whipping}} \text{ X } 100$$

2.9. Differential scanning calorimetric (DSC) analysis and thermal gravimetric analysis (TGA)

Differential scanning calorimetry (DSC) was carried out using a Mettler Toledo Star SW 7.01, according to the procedure of Meng & Ma (1981), with slight modification. The protein sample (5 mg) was dissolved in 1 ml of 0.06 M phosphate buffer (pH 7.0) containing 0.1 M NaCl. A 45 µl of protein solution was hermetically sealed in a stainless steel pan, and was heated from 0 to 300°C at a rate of 10 °C/min. The thermal properties were referenced against another pan containing 45 mL of buffer without protein. The denaturation peak temperature (dT) and enthalpy (DH) were calculated by a thermal analysis software program. The temperature at which denaturation started, known as the onset denaturation temperature "T onset", was calculated by taking the intercept of the baseline and the extrapolated maximum slope of the peak. The peak denaturation temperature "T peak" was considered to be the temperature at maximum heat flow. The enthalpy of thermal denaturation was calculated from the area of the endothermic peak. Thermal gravimetric analysis (TGA) was carried with Mettler

ToledoTGA/SDTA System (Greifensee, Switzerland), and the thermogram was obtained in the range of 30 – 480 °C at a rate of 10 °C/min.

2.10. FT-IR Spectroscopy

The lyophilized protein concentrate was ground with potassium bromide at a 1/100 ratio (w/w). This protein concentrate was pressed at high pressure into a KBr pellet. The spectral analysis was carried out using NXR FT-IR module (Thermo electron corporation USA). The FT–IR spectrum of sample was recorded in the 4000-400 cm^{-1} region at room temperature.

2.11. Statistical analysis

Analysis of variance (ANOVA) was conducted to determine the differences among all treatments and multiple comparison tests with the least significance difference (LSD) were performed to determine significant differences.

3. Result and Discussion

3.1. Protein content, recovery and yield of *K. alvarezii*

K. alvarezii biomass constituted 18.16 ± 0.03 % total protein determined on dry weight which is analogous to the report of Rajasulochan, Krishnamoorthy & Dhamotharan (2012) who reported 18.78 % protein content for the same species. The protein content of *K. alvarezii* obtained herein was comparable or in fact higher than that reported for seaweeds such as *Caulerpa veravelensis* (7.77 ± 0.59 %), *C. scalpelliformis* (10.50 ± 0.91 %) and *C. racemosa* (12.88 ± 1.17 %), *Plocamium brasiliense* (15.72%), *Ochtodes secundiramea* (10.10%,), *Laminaria japonica* (9.1%), *Palmaria palmate* (13.5%) and *Ulva rigida* (17.8%) (Kumar, Gupta, Kumari, Reddy & Jha, 2011; Gressler, Fujii, Martins, Colepicolo, Mancini-Filho & Pintoa, 2011).

Using ammonium sulfate precipitation method, 7.81 ± 2.42 % of protein concentrate could be recovered; this PC was comprised of 62.3 ± 1.62 % protein. The protein content of the PC was much higher than that reported by Ganesan, Kumar & Subba Rao (2012); they studied total protein content and protein contents in PC of *Enteromorpha tubulosa* (19.09 ± 0.91; 53.83 ± 0.70%), *Enteromorpha compressa* (17.48 ± 0.41; 60.35 ± 2.01%) and *Enteromorpha linza* (12.5 ± 1.26; 33.36 ±1.04 %) and could recover up to 6.16, 6.48 and 5.71 % of PC respectively with each of the respected species. The figures were also higher than that obtained by Wong & Cheung (2001) who studied total protein content and protein content of the PC of *Hypnea charoides* (18.13 ± 0.29; 55.4 ± 0.63 % respectively), *Hypnea japonica* (19.4 ± 0.33; 56.67 ± 0.25 %) and *Ulva lactuca* (7.13 ± 0.21; 50.87 ± 0.5%).

3.2. Nitrogen Solubility

The effect of pH and salt concentration on nitrogen solubility of *K. alvarezii* PC is elaborated in Fig. 1. The minimum nitrogen solubility (33.72 ± 1.23 %) was observed at pH 4; this might be due to the fact that the isoelectric point (pI) of the protein might be corresponding here. Sorgentini & Wagner (2002) suggest that majority of the food proteins being acidic proteins exhibit minimum solubility at pH 4 – 5 and maximum solubility at alkaline pH. No much difference in nitrogen solubility values were noticed at pH 8 and 10. The PC showed only a gradual increase in solubility from pH 8–12 in water as well as NaCl concentrations. However, at pH 12, *K. alvarezii* PC reached 58.72 ± 1.68 % solubility in the presence of 0.5 M NaCl, which was comparatively lower than PC of *Enteromorpha tubulosa* (13.60 ± 0.85 to 25.41 ± 1.94%), *Enteromorpha compressa* (14.96 ± 0.35 to 26.38 ± 0.88%) and *Enteromorpha linza* (10.87 ± 1.12 to 20.31 ± 1.66%) (Ganesan, Kumar & Subba Rao, 2012); it was also lower than that reported for fenugreek PC (86.3% at pH 10) (El Nasri & Tinay, 2007). According to Seena & Sridhar (2005), at highly acidic and alkaline pH, the protein acquires a net positive and negative charge respectively which favors the repulsion of molecules and thereby increases the solubility of the protein. The solubility is a physico-chemical property of a protein that critically affects its functional properties as manifested in foods, mainly emulsifying, foaming, and gel forming abilities. On this accord, *K. alvarezii* PC demonstrated remarkable solubility values in the presence of salt at varying pH it could be stated that this PC would definitely hold a promising market in various food products.

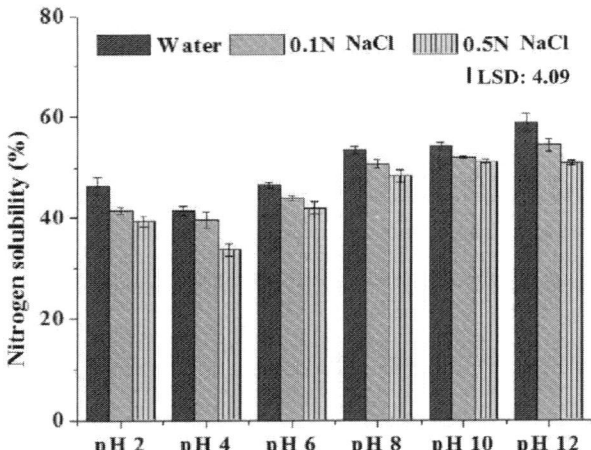

Fig. 1. Influence of pH and ionic strength contributed by NaCl on nitrogen solubility of *K. alvarezii* protein concentrate.

3.3. Water-holding and fat absorption capacity (WHC and FAC)

The water-holding capacity of *K. alvarezii* PC was 2.22 ± 0.04 g water/g of PC; this was higher than that reported for *E. compressa* (1.53 ± 0.07 g water/g PC), *E. tubulosa* (1.32 ± 0.11 g water/g PC) and *E. linza* (1.22 ± 0.06 g water/g PC) (Ganesan, Kumar & Subba Rao, 2012); however, this value was lower than that reported for the protein concentrate of Egyptian fenugreek (3.52 ml water/g of protein) (Abdel–Aal, Yousif, Adel-Shehata & El–Mahdy, 1986). High water absorption of proteins helps to reduce moisture loss in packaged bakery goods; moreover water-holding is indispensably required to maintain freshness and moist mouth feel of baked foods. The water-holding capacity is a critical property of proteins in viscous foods, e.g. soups, dough, custards and baked products, because these are supposed to imbibe water without dissolution of protein, thereby providing body, thickening and viscosity (Cooper & Goldenberg, 1987; Ganesan, Kumar & Subba Rao, 2012). However, the intrinsic factors affecting the water–binding capacity of food proteins include amino acid composition, protein conformation, and surface polarity/hydrophobicity. The very fact that *K. alvarezii* PC holds tremendous water-holding capacity, suggests its appropriateness in being used in various delicacies requiring moist foods.

The fat absorption capacity of *K. alvarezii* PC was 1.29 ± 0.20 g oil/g of protein was slightly lower than that of *E. compressa* (1.34 ± 0.10 g oil/g PC) and fenugreek (*Trigonella foenum graecum*) protein concentrate (1.56 g oil/g PC); but it was slightly higher than that reported for *E. linza* (1.05 ± 0.07 g oil/g PC) and *E. tubulosa* (1.08 ± 0.04 ml oil/g PC) (Ganesan, Kumar & Subba Rao, 2012; El Nasri & Tinay, 2007). In food systems, good interactions of water and oil with proteins are imperative as this would indirectly affect the flavor and texture of foods. However, food processing methods have important impacts on the protein conformation and hydrophobicity. The fat/oil holding/absorption capacity is a critical determinant of flavor retention, while fat emulsion capacity and stability are important attributes of additives for the stabilization of fat emulsions. High oil absorption/holding is requisite for the formulation foods such as sausages, cake batters, mayonnaise, and salad dressings (Chandi & Sogi, 2007).

As the PC of *K. alvarezii* demonstrated acceptable water-holding and oil-holding capacity, it could be definitely be used for multiple food applications such as water-holding, or as a texture enhancer. It could also be used suitable for improving the viscous nature of food formulations.

3.4. Emulsifying and surface active properties

Emulsifying properties of food proteins can be described by the emulsifying activity (EAI) and emulsifying stability (ESI) indices. The emulsifying activity (EAI) is a measure of the capacity of the protein to help formation and stabilization of the emulsion created while ESI provides a measure of the ability of the protein to impart strength to the emulsion to resist changes to its structure (e.g., coalescence, creaming, flocculation or sedimentation) over a defined time period (Boye, Aksay, Roufik, Ribéreau, Mondor, Farnworth & Rajamohamed, 2010). *K. alvarezii* PC efficiently emulsified oils such as silicone oil, paraffin oil, groundnut oil, cotton seed oil, Jatropha oil, cedar wood oil, jojoba oil, sunflower oil and olive oil. The emulsification indices of *K. alvarezii* PC with various oils have been shown in Fig. 2; here, maximum emulsification indices were observed with cedar wood oil (99.67 ± 0.58), Jatropha oil (99.33 ± 1.15) and olive oil (99 ± 1.73) after 15 min. Moreover, this PC showed good emulsifying activity with groundnut oil (77 ± 1.00) and cotton seed oil (75.68 ± 0.58) after 15 min. Formation of stable emulsions was observed using cedar wood oil (E_{720} = 75.33 ± 2.08), olive oil (E_{720} = 54.33 ± 1.16) and Jatropha oil (E_{720} = 53.67 ± 1.59) at 10 mg PC/ml concentration, the emulsion

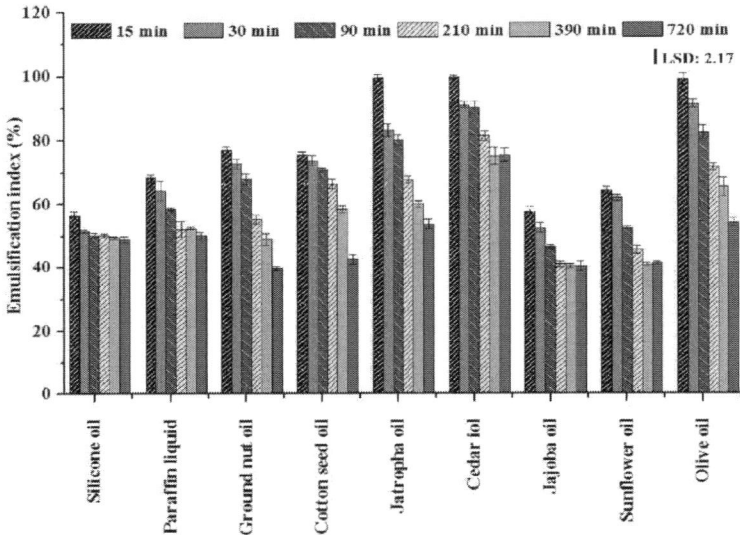

Fig. 2. Emulsifying index of *K. alvarezii* protein concentrate using different oils and emulsion stability.

stability of *K. alvarezii* PC could probably depend on the strength of the hydrophilic and hydrophobic properties of proteins rather than the balance between hydrophilic and lipophilic phases. Reports suggest that the hydrophobic lipid portions in an emulsion are generally responsible for emulsifying action (Gbadamosi, Abiose & Aluko, 2012). In this study, the *K. alvarezii* PC not only formed emulsions with superior emulsifying indices but demonstrated stable emulsions with various oils including edible oils, this is extremely essential for its applicability as an emulsifier. Dickinson & Galazka (1991) suggest that the emulsifying activity of products often depend to a great extent on the nature and concentration of the protein present in it (for e.g. acacia gums). Moreover, a high percentage of hydrophobic amino acids in the protein moiety favor emulsification (Dickinson & Galazka, 1991). Thus in the present study too the amount of protein present in the concentrate might be responsible for the formation of stable emulsions.

The surface tension of distilled water (which was used as a reference or control, i.e. without any surface-active agent) was 72.05 ± 0.04 mN.m^{-1}, while the surface tension of 0.1 and 0.5% of *K. alvarezii* PC was 50.10 ± 0.03 and 44.02 ± 0.03 mN/m respectively; this was comparable with milk protein in various products such as Prolacta 90, Promilk 852 FB, Promilk 852 B, skimmed milk powder as well as sodium caseinate was 46.1, 46.2, 49.4, 48.4 and 46.4 respectively. Emulsifiers are most effective and stable in lowering surface tension (Rouimi, Schorsch, Valentini & Vaslin, 2005); therefore the very fact that the *K. alvarezii* PC was surface active and could lower the surface tension of distilled water, qualifies it as a emulsifying agent. There are several reports available on the use of several biological products of animal, plant and bacterial origin for the reduction of surface tension, however, the surface tension properties of seaweed protein has barely been dealt with till date.

3.5. Foaming capacity (FC) and stability (FS)

The foaming capacity (FC) of PC investigated herein was pH-dependent (Fig. 3); lowest FC (38 ± 2%) was recorded at pH 6.0. On the other hand highest FC (53.33 ± 2.31%) was obtained at pH 4.0; this value was slightly lower than that of *E. compressa* (55.0 ± 2.6%; Ganesan, Kumar & Subba Rao, 2012), and much lower than the fenugreek protein concentrate i.e. 89.5 % (El Nasri & Tinay, 2007). It was much higher than that of *E. tubulosa* (31.9 ± 2.7%) and *E. linza* (33.3 ± 5.7%) (Ganesan, Kumar & Subba Rao, 2012). Here, maximum foaming stability (FS) (45.33 ± 1.15) recorded after 30 min at pH 2.0 was higher than *E. compressa* (37.5 ± 2.0%), *E. tubulosa* (16.7 ± 1.5%) and *E. linza* (4.4 ± 2.0) (Ganesan, Kumar & Subba Rao, 2012).

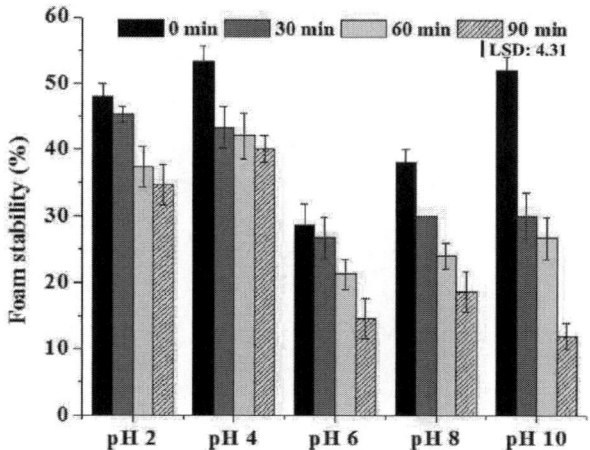

Fig. 3. Effect of pH and time on the foaming stability of *K. alvarezii* protein concentrate.

The basic requirements for a protein to be a good foaming agent are the ability to adsorb rapidly at the air-water interface during bubbling; and the ability to undergo rapid conformational changes at the interface. Furthermore, the high foaming ability mainly depended on protein dispersing ability, but stability of foaming was primarily influenced by the degree of denatured protein (Fidantsi & Doxastakis, 2001).

3.6. Differential scanning calorimetric (DSC) analysis and thermal gravimetric analysis (TGA)

It is known that formation of unique structures of biological macromolecules, such as proteins and their specific complexes is, in principle, reversible, and the reactions are thermodynamically driven; therefore, thermodynamic investigations of these processes are of high priority (Gill, Moghadam & Ranjbar, 2010). DSC is a rapid, easy, and capable technique for supplying both thermodynamic (heat capacity, enthalpy, and entropy) and kinetic data (reaction rate and activation energy) on protein denaturation, and has been used extensively in various food systems. The information on protein thermal properties is useful for food-processing strategies and heat-processing design (Ju, Hettiarachchy & Rath, 2001). Fig. 4 shows typical DSC thermograms of *K. alvarezii* protein concentrate in 0.06M phosphate buffer (pH 7.0). The protein sample was heated from 0 to 300 °C at 10 °C /min. The PC exhibited two observable endothermic peaks; the minor endothermic peak temperature (T_m) at about 108.52 °C and the major one at 109.25 °C (T_M), the enthalpy of the thermal denaturation was ΔH - 5.3 J/mg. It has been demonstrated that the enthalpy (ΔH) reflects the status of ordered conformation of food proteins (Koshiyama, Hamano, & Fukushima, 1981). Therefore, the "net" ΔH indicates cumulative effects of endothermic events such as the breakdown of hydrogen bonds and exothermic phenomena such as aggregation of food proteins due to hydrophobic interactions (Murray, Arntfield, & Ismond, 1985). The appearance of minor endothermic peak shoulder might be due to the presence of carrageenan and the second big transition peak presumably represents the denaturation of high percentage of *K. alvarezii* PC. A similar trend has been observed in case of a mixture of whey protein isolate (WPI) and the

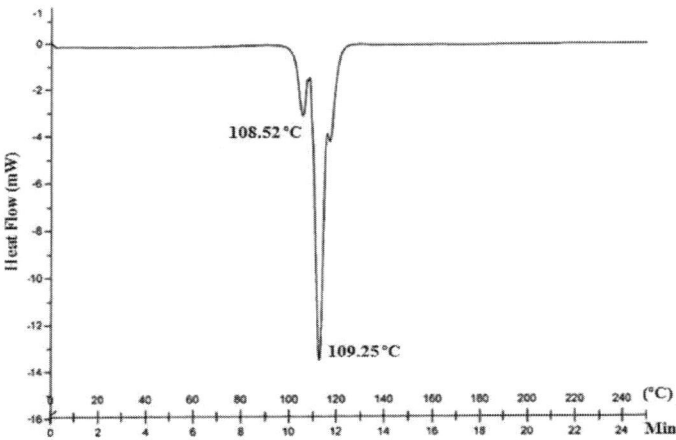

Fig. 4. Typical DSC thermograms of *K. alvarezii* protein concentrate.

hydrocolloid *i*-carrageenan, where the thermogram showed two transition peaks at 52 and 78.5 °C; however, in that report the first transition peak observed in the WPI + *i*-carrageenan mixture appeared in the presence of carrageenan while the second big transition peak in both thermograms probably represented the denaturation of high percentage of this protein (Ibanoglu, 2005).

TGA carried out dynamically between weight loss vs. temperatures have been elucidated in Fig. 5. TGA showed that degradation of *K. alvarezii* protein concentrate takes place in two steps: here, the 5.4% weight loss of total PC was recorded from 30 to 100 °C which could be due to moisture content, thereafter a second phase of degradation (43.5%) was observed with maximum weight loss at 260 °C. However, the total weight loss of PC occurred on further increase of temperature. This is chiefly associated with degradation of the major protein component of *K. alvarezii* PC.

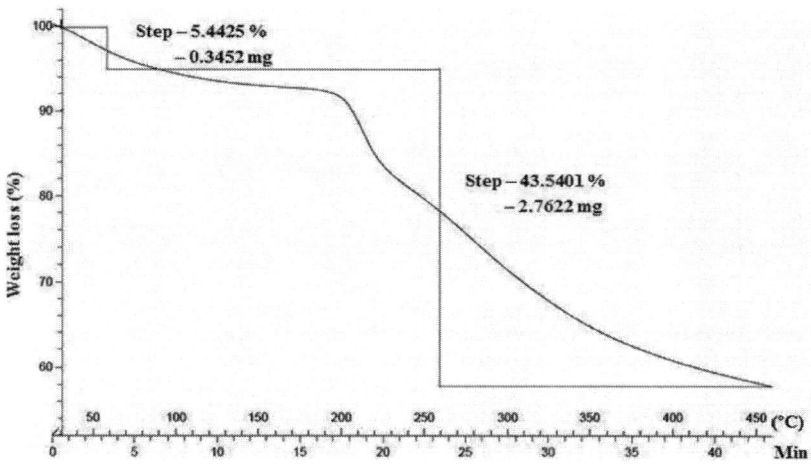

Fig. 5. Typical TGA thermogram of *K. alvarezii* protein concentrate.

3.7. FT-IR spectroscopy

Proteins are frequently referred to as having a certain fraction of structural components (α-helix, β-sheet, etc.). The secondary structural composition is some of the most important information for a structure-unknown protein. Therefore estimation of protein secondary structure is one of the major applications of the FT–IR technique (Kong & Yu, 2007). FT–IR spectra provide information about the structural composition of proteins. The spectrum of the PC (Fig. 6) showed a band at 616 /cm which could be due the presence of phosphate group, a stretching band at 704 /cm, reveals out of plane N–H bending (El-Bahy, 2005; Jung, Stuehr & Ghosh, 2000). Kumar & Kaladharan (2007) discretely describe *K. alvarezii* to contain numerous amino acids including 0.08% histidine; however, they further emphasize on the utility of seaweeds as a potential low-cost source of protein for fish.

73

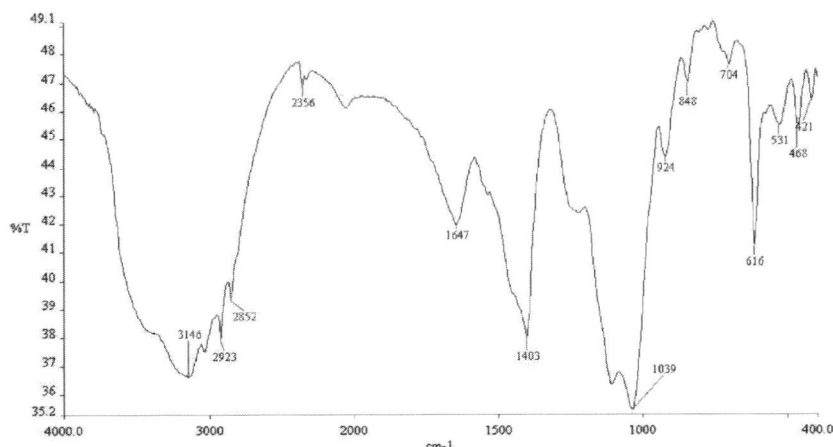

Fig. 6. FT-IR Spectrum of *K. alvarezii* protein concentrate.

4. Conclusion

Protein concentrates and isolates prevalently used in food industry are mainly derived from dairy, soy or wheat; however, certain reports suggest that these could trigger allergic responses. The nutritional quality of plant proteins is lower than animal protein, and, plant PCs could at times possess anti-nutritional factors; therefore, food manufacturers as well as consumers are looking for alternative protein sources which could be economically feasible and available all year round. Keeping in mind that seaweeds have been used as traditional food since ages, it could be stated that seaweeds are probably the best alternative in this regard. It should also be noted that in the seaweed context, data regarding anti-nutritional factors are virtually non–existent. In addition to providing nutrition, it is necessary that the proteins being used in food formulation should possess specific functional properties that facilitate processing and serve as the basis of product performance and it helps tailoring them to design food products of our interest. Thus, in order to demonstrate the applicability of seaweed PCs in the food formulations, in-depth studies need to be carried out in this regard. On this consensus, the *K. alvarezii* protein concentrate studied herein demonstrated admirable functional properties at par with most other protein concentrates; furthermore, its efficacy to be used at varying pH ranges and in the presence of salt indeed qualifies this PC to be incorporated into several value-added food products.

References

Abdel-Aal, E.S.M., Yousif, M.M., Adel-Shehata, A., & El-Mahdy, A.R. (1986). Chemical and functional properties of some legumes powder. *Egyptian Journal of Food Science, 13,* 201–205.

Bencini, M.C. (1986). Functional properties of drum dried chickpea (*Cicer arietinum* L.) flour. *Journal of Food Science, 51,* 1518–1526.

Bera, M.B., & Mukherjee, R.K. (1989). Solubility, emulsifying, and foaming properties of rice bran protein concentrates. *Journal of Food Science, 54,* 142–145.

Boye, J.I., Aksay, S., Roufik, S., Ribéreau, S., Mondor, M., Farnworth, E., & Rajamohamed, S.H. (2010). Comparison of the functional properties of pea, chickpea and lentil protein concentrates processed using ultrafiltration and isoelectric precipitation techniques. *Food Research International, 43,* 537–546.

Chandi, G.K., & Sogi, D.S. (2007) Functional properties of rice bran protein concentrates. *Journal of Food Engineering, 79,* 592–597.

Cooper, D.G., & Goldenberg, B.G. (1987). Surface-active agents of two *Bacillus* species. *Applied and Environmental Microbiology, 53,* 224–229.

Dickinson, E., Galazka, V.B., & Anderson, D.M.W. (1991). Emulsifying behavior of gum Arabic. Part I—effect of the nature of the oil phase on the emulsion droplets. *Carbohydrate Polymer, 14,* 373–383.

El Nasri, A.H., El Tinay, A.H. (2007). Functional properties of fenugreek (*Trigonella foenum graecum*) protein concentrate. Food Chemistry, 103, 582–589.

El-Bahy, G.M.S. (2005). FTIR and Raman spectroscopic study of Fenugreek (*Trigonella foenum graecum* L.) seeds. *Journal of Applied Spectroscopy, 72,* 111–116.

Fidantsi, A., & Doxastakis, G. (2001). Emulsifying and foaming properties of Amaranth seed protein isolate. *Journal of Colloids and Surfaces, 21,* 119–124.

Fleurence, J. (1999) Seaweed proteins: biochemical, nutritional aspects and potential uses. *Trends in food science and technology, 10,* 25–28.

Fleurence, J., Le Coeur, C., Mabeau, S., Maurice, M., & Landrein, A. (1995). Comparison of different extractive procedures for proteins from the edible seaweeds *Ulva rigida* and *Ulva rotundata*. *Journal of Applied Phycology, 7,* 577–582.

Ganesan, K., Kumar, K.S., & Subba Rao, P.V. (2012). Salt- and pH-induced functional changes in protein concentrate of edible green seaweed *Enteromorpha* species. *Fisheries science, 78,* 169–176.

Gbadamosi, S.O., Abiose, S.H., & Aluko, R.E. (2012). Amino acid profile, protein digestibility, thermal and functional properties of Conophor nut (*Tetracarpidium conophorum*) defatted flour, protein concentrate and isolates. *International Journal of Food Science and Technology, 47,* 731–739.

Gill, P., Moghadam, T.T., & Ranjbar, B. (2010). Differential Scanning Calorimetry Techniques: Applications in Biology and Nanoscience. *Journal of Biomolecular Techniques, 21,* 167–193.

Gressler, V., Fujii, M.T., Martins, A.P., Colepicolo, P., Mancini-Filho, J., & Pintoa, E. (2011). Biochemical composition of two red seaweed species grown on the Brazilian coast. *Journal of the Science of Food and Agriculture, 91,* 1687–1692.

Ibanoglu, E. (2005). Effect of hydrocolloids on the thermal denaturation of proteins. *Food Chemistry, 90,* 621–626.

Ju, Z.Y., Hettiarachchy, N.S., & Rath, N. (2001). Extraction, denaturation and hydrophobic properties of rice flour proteins. *Journal of food science, 66,* 229–232.

Jung, C., Stuehr, D.J., & Ghosh, D.K. (2000). FT-infrared spectroscopic studies of the iron ligand CO stretch mode of iNOS oxygenase domain: Effect of arginine and tetrahydrobiopterin. *Biochemistry, 39,* 10163–10171.

Kong, J., & Yu, S. (2007). Fourier Transform Infrared Spectroscopic Analysis of Protein Secondary Structures. *Acta Biochimica et Biophysica Sinica, 39(8),* 549–559.

Koshiyama, I., Hamano, M., & Fukushima, D. (1981). A heat denaturation study of the 11S globulin in soybean seeds. *Food Chemistry, 6(4),* 309–322.

Kumar V.V. & Kaladharan P. (2007). Amino acids in the seaweeds as an alternate source of protein for animal feed. *Journal of the Marine Biological Association of India, 49(1),* 35–40.

Kumar, M., Gupta, V., Kumari, P., Reddy, C.R.K., & Jha, B. (2011). Assessment of nutrient composition and antioxidant potential of Caulerpaceae seaweeds. *Journal of Food Composition and Analysis, 24,* 270–278.

Meng, G.T., & Ma, C.Y. (2001). Thermal properties of *Phaseolus angularis* (red bean) globulin. *Food Chemistry, 23,* 453-460.

Murray, E.D., Arntfield, S.D., & Ismond, M.A.H. (1985). The influence of processing parameters on food protein functionality. II. Factors affecting thermal properties as analyzed by differential scanning calorimetry. *Canadian Institute of Food Science and Technology Journal, 18(2),* 158–162.

Nath, JP., & Rao, M.S.N. (1981). Functional Properties of Guar Proteins. *Journal of Food Science, 46,* 1255–1259.

Ogunwolu, S.O., Henshaw, F.O., Mock, H.P., Santros, A., & Awonorin, S.O. (2009). Functional properties of protein concentrates and isolates produced from cashew (*Anacardium occidentale* L.) nut. *Food Chemistry, 115,* 852–858.

Rajasulochana, P., Krishnamoorthy, P., & Dhamotharan, R. (2012). Biochemical investigation on red algae family of *Kappahycus* sp. *Journal of Chemical and Pharmaceutical Research, 4(10),* 4637-4641.

Rouimi, S., Schorsch, C., Valentini, C., & Vaslin, S. (2005). Foam stability and interfacial properties of milk protein–surfactant systems. *Food Hydrocolloids, 19,* 467–478.

Seena, S., & Sridhar, K.R. (2005). Physiochemical, functional and cooking properties of Canavalia. *Journal of Food Chemistry, 32,* 406–412.

Slaski, R.J., & Franklin, P.T. (2011). A review of the status of the use and potential to use micro and macroalgae as commercially viable raw material sources for aquaculture diets. Report commissioned by Scottish Aquaculture Research Forum (SARF), 94 pp. Available from: http://www.sarf.org.uk/cms-assets/documents/29524-222388.sarf077.pdf

Sorgentini, D.A., & Wagner, J.R. (2002). Comparative study of foaming properties of whey and isolate soybean proteins. *Food Research International, 35,* 721–729.

Wathelet, B., (1999). Nutritional analyses of proteins and amino acids in beans (*Phaseolus* sp.). *Biotechnology, Agronomy, Society and Environment 3,* 197–200.

Wong, K.H., & Cheung, P.C.K. (2001). Nutritional evaluation of some subtropical red and green seaweeds Part II. In vitro protein digestibility and amino acid profiles of protein concentrates. *Food Chemistry, 72,* 11–17.

Wong, K.H., Cheung, P.C.K., & PO Ang Jr. (2004), Nutritional evaluation of protein concentrates isolated from two red seaweeds: *Hypnea charoides* and *Hypnea japonica* in growing rats. *Hydrobiology, 173,* 271-278.

Yun, J.H., Kwon, I.K., Lohakare, J.D., Choi, J.Y., Yong, J.S., Zheng, J., Cho, W.T., & Chae. B.J. (2005). Comparative efficacy of plant and animal protein sources on the growth performance, nutrient digestibility, morphology and caecal microbiology of early-weaned pigs. *Asian - Australasian Journal of Animal Sciences, 18,* 1285–1293.

Chapter IV
Antioxidant potential of Kappaphycus alvarezii

1. Introduction

All living organisms contain complex systems of antioxidant enzymes and chemicals. Some of these systems, like the thioredoxin system, are conserved through out evolution and are required for life. Antioxidants in biological systems have multiple functions, including defending against oxidative damage and participating in the major signaling pathways of the cells. One major action of antioxidants in cells is to prevent damage caused by the action of reactive oxygen species. Reactive oxygen species include hydrogen peroxide (H_2O_2), the superoxide anion (O_2^-), and free radicals such as the hydroxyl radical ($\cdot OH$). These molecules are unstable and highly reactive, and can damage cells by chemical chain reactions such as lipid peroxidation, or formation of DNA adducts that could cause cancer–promoting mutations or cell death. In order to reduce or prevent this damage all cells invariably contain antioxidants.

Lipid oxidation by reactive oxygen species (ROS) such as superoxide anion, hydroxyl radicals, and hydrogen peroxide causes a decrease in nutritional value of lipids, in their safety and appearance. In addition, it is the predominant cause of qualitative decay of foods, which leads to rancidity, toxicity, and destruction of biochemical components important in physiologic metabolism. Free radicals–mediated modification of DNA, proteins, lipids, and small cellular molecules are associated with a number of pathological processes, including atherosclerosis, arthritis, diabetes, cataractogenesis, muscular dystrophy, pulmonary dysfunction, inflammatory disorders, ischemiareperfusion tissue damage, and neurological disorders such like Alzheimer's disease (Frlich & Riederer,1995).

Antioxidants are classified by the products they form on oxidation (these can be antioxidants themselves, inert, or pro-oxidant), by what happens to the oxidation products (the antioxidant may be regenerated by different antioxidants or, in the case of "sacrificial" antioxidants, its oxidized form may be broken down by the organism) and how effective the antioxidant is against specific free radicals. Several synthetic antioxidants such as butylated hydroxyanisole (BHA), butylated hydroxytoluene (BHT), and butylated hydroxyquinone (TBHQ) are commercially available and currently used. However, these antioxidants have been restricted for use in foods as they are suspected to be carcinogenic. Some toxicological studies have also implicated the use of these synthetic antioxidants in promoting the development of cancerous cells in rats. These findings, together with consumers' interests in natural food additives, have reinforced the efforts for the development of alternative antioxidants from natural origins (Huang & Wang, 2004). An immense number of marine flora and fauna are reported to have wide spectrum of interesting biological properties. In folk medicines seaweeds have been used for a variety of remedial purposes for the treatment of eczema, gallstone, gout, crofula, cooling agent for fever, menstrual trouble, renal problems, scabies, etc. (Chapman & Chapman, 1976).

Seaweeds are rich in polysaccharides, minerals, proteins and vitamins. Documented antioxidant activity would elevate their value in the human diet as food and pharmaceutical supplements (Yan, Nagata & Fan, 1998). Few reports are available on the antioxidant potential of seaweeds (Jimenez–Escrig, Jimenez–Jimenez, Pulido & Saura–Calixto, 2001). Ismail & Hong (2002) reported antioxidant activity of four commercial edible seaweeds namely Nori (*Porphyra* sp.), Kumbu (*Laminaria* sp.), Wakame (*Undaria* sp.) and Hijiki (*Hijikia* sp.).

The Rhodophyta (red algae) are a distinct eukaryotic lineage characterized by the accessory photosynthetic pigments phycoerythrin, phycocyanin and allophycocyanins arranged in phycobilisomes. They contain a large assemblage of species that predominant the coastal and continental shelf areas of tropical, temperate and cold–water regions. Red algae are ecologically significant as primary producers, providers of structural habitat for other marine organisms, and they play an important role in the primary establishment and maintenance of coral reefs. Some red algae are economically important as providers of food and gels (Wilson 2000). For this reason, extensive farming and natural harvest of red algae occurs in numerous areas of the world. *Kappaphycus alvarezii,* an economically important red tropical seaweed which is highly demanded for its cell wall polysaccharides, is the most important source of *kappa* carrageenan. The world production of *Kappaphycus sp.* is approximately 28000 tons per annum. This seaweed accounts for the largest consumption worldwide (Mc Hugh, 1987). It is easily accessible in huge amounts for food and pharmaceutical applications. The present study deals with antioxidant properties of *K. alvarezii*.

2. Materials and Methods

2.1. Collection of samples

Kappaphycus alvarezii was collected from a cultivation site at Port Okha (L 22° 28.528' N; L 069° 04.322' E) located on the North West coast of India during April 2006. The sample was thoroughly washed with seawater to remove epiphytes and dirt particles followed by shade drying for two days. It was then brought to the laboratory, oven dried at 70 $^\circ$C for 4 h to obtain a constant weight and pulverized in the grinder (size 2 mm). This sample was used for determination of phenolic content as well as antioxidant studies. The chemicals used in these studies were of analytical grade.

2.2. Preparation of Extracts

The pulverized moisture free sample (20 g) was extracted with 200 ml of individual solvents using a Soxhlet extractor. The extraction was repeated many times to obtain a sizable quantity of extract. Consequently, the extract was concentrated in a rotary evaporator at 40°C. Different solvents were used for the preparation of extracts to determine the antioxidant efficacy of *K. alvarezii*. All the experiments were conducted in triplicate.

2.3. Determination of total phenol

Total phenolic content was estimated by Folin–Ciocalteau method (Singleton & Rossi, 1965). To 6.0ml double distilled water, 0.1 ml sample and 0.5 ml Folin–Ciocalteau reagent was mixed followed by the addition of 1.5 ml Na_2CO_3 (20 g 100 ml^{-1} water) and the volume was made up to 10.0 ml with distilled water. After incubation for 30 min at 25 °C, the absorbance was measured at 760 nm and the total phenolic content was calculated with gallic acid standard and expressed as percentage of total phenols obtained on dry weight basis.

2.4. DPPH Radical Scavenging Assay

DPPH scavenging potential of different fractions was measured based on scavenging ability of stable 1, 1–diphenyl–2-picrylhydrazyl (DPPH) radicals by *K. alvarezii* antioxidants. The ability of extracts to scavenge DPPH radicals was determined according to the method of Blois (1958). Briefly, 1 ml of 1 mM methanolic solution of DPPH was mixed with 1 ml of extract solution (containing 0.5-5.0 mg ml^{-1} of dried extract). The mixture was then vortexed vigorously and left for 30 min at room temperature in the dark. The absorbance was measured at 517 nm and activity was expressed as percentage DPPH scavenging activity relative to control using the following equation:

$$\% \text{ Radical scavenging activity} = [A_{Control} - A_{Sample} / A_{Control}] \times 100$$

2.5. Ferrous ion chelating activity

Iron chelating ability of methanol, ethanol and ethyl acetate extracts of *K. alvarezii* were used for the present investigation. The chelating of ferrous ions by the extracts and standards was estimated by the method of Dinis, Madeira & Almeida (1994). Extracts were added to a solution of 2mM $FeCl_2$ (0.05 ml). The reaction was initiated by the addition of 5mM ferrozine (0.2 ml) and the mixture was shaken vigorously and left idles at room temperature for 10 min. After the mixture had reached equilibrium, the absorbance of the solution was then measured at 562 nm. The percentage of inhibition of ferrozine $-Fe^{2+}$ complex formation was determined using the following formula:

$$\% \text{ Inhibition} = [1 - A_{Sample} / A_{Control}] \times 100$$

Where A_0 was the absorbance of the control and A_1 was the absorbance in the presence of the sample extracts and standards. The control contains $FeCl_2$ and ferrozine, with complex formation molecules.

2.6. Reducing power

Extracts of *K. alvarezii* were prepared using methanol, ethanol, water, ethyl acetate and hexane. The reductive potential of extracts was determined according to the method of Oyaizu (1986). The different concentrations of extracts (0.5-25 mg ml^{-1}) was mixed with phosphate buffer (2.5 ml, 0.2 M, pH 6.6) and potassium ferricyanide [$K_3Fe(CN)_6$] (2.5 ml, 1%). The mixture was incubated at 50 °C for 20 min. A portion (2.5 ml) of trichloroacetic acid (10%) was added to the mixture, which was then subjected to centrifugation (10 min, 1000g). The upper layer of solution (2.5 ml) was mixed with distilled water (2.5 ml) and $FeCl_3$ (0.5 ml, 0.1%), and the absorbance was measured at 700 nm. Higher absorbance of the reaction mixture indicated greater reductive potential.

2.7. Antioxidant activity in linoleic acid system with ferrothiocyanate reagent (FTC)

Ethanolic extract of *K. alvarezii* was subjected to the assay adopted by Osawa and Namaki (1983). The extract (4 mg) was dissolved in 99.5% ethanol and mixed with 2.5% linoleic acid in 99.5% ethanol (4.1 ml), 0.05 M phosphate buffer (pH = 7, 8 ml) and distilled water (3.9 ml) and kept in screw cap containers under dark conditions at 40 °C; 0.1 ml of this solution was added to 9.7 ml of 75% ethanol and 0.1 ml of 30% ammonium thiocyanate. After 3 min, 0.1 ml of 0.02 M ferrous chloride in 3.5% hydrochloric acid was added to the reaction mixture, the absorbance of red colour was measured at 500 nm in the spectrophotometer, for every two days. The control and standard were subjected to the same procedure except for the control, where there was no addition of sample and for the standard 4 mg of sample were replaced with 4 mg of Butylated hydroxy toluene (BHT) used as a positive control. Absorbance was measured at an interval of 2 days. The percent inhibition of linoleic acid peroxidation was calculated as:

Inhibition (%) = 100 - [(absorbance increase of the sample/absorbance increase of the   control) × 100].

The IC_{50} value represented the concentration of the compounds that caused 50% inhibition. All experiments were carried out in triplicate.

2.8. Statistical analysis

For the extract, three samples were prepared for each experiment. The data were presented as mean ± standard deviation.

3. Results and Discussion

3.1. Antioxidant activity

The antioxidant activity is system dependent. Moreover it depends on the method adopted and the lipid system used as substrate (Singh, Maurya, de Lampasona, & Catalan, 2006). Hence, different following methods have been adopted in order to assess the antioxidative potential of *K. alvarezii* extracts.

3.2. Total phenol content

A number of studies have focused on the biological activities of phenolic compounds, which are potential antioxidants and free radical scavengers (Rice-Evans, Miller, Bolwell, Bramley & Pridham, 1995; Marja, Ka¨hko¨nen, Anu, Hopia, Heikki, Vuorela, Rauha, Pihlaja, Kujala & Heinonen, 1999; Sugihara, Arakawa, Ohnishi, & Furuno, 1999). The total phenol content was maximum when a mixture of chloroform and methanol (2:1) was used (2.048 ± 0.038 %) followed by ethanol (1.939 ± 0.029 %), methanol (1.793 ± 0.077 %), n-propanol (1.404 ± 0.040 %) and ethyl acetate (1.091 ± 0.597 %). Extracts obtained using other solvents, viz. acetone, n-hexane and chloroform, showed < 1% total phenol content (Table 1).

Table 1
Percent phenol content of *K. alvarezii* in various solvents

Solvents	Total phenol (%)
Acetone	0.963 ± 0.058
n-Propanol	1.40 ± 0.040
Ethyl acetate	1.09 ± 0.597
n-Hexane	0.83 ± 0.048
Chloroform	0.683 ± 0.040
Methanol	1.79 ± 0.77
Ethanol	1.94 ± 0.029
Chloroform:methanol (2:1)	2.05 ± 0.038

Values are means of three replicate determinations; SD, standard deviation.

3.3. Scavenging effect on 1,1–diphenyl–2–picrylhydrazyl radical (DPPH)

The 1, 1–diphenyl–2–picrylhydrazyl (DPPH) radical is a stable radical with a maximum absorbance at 517 nm that can readily undergo reduction by an antioxidant. Because of the ease and convenience of this reaction, it has now widespread use in the free radical scavenging activity assessment (Brand-Williams, Cuvelier, & Benset, 1995). The radical scavenging activity of *K. alvarezii* extract is shown in Figure 1 and expressed as percentage reduction of the initial DPPH absorption by the tested compound. The best radical scavenging activity could be obtained in the ethanol extract (IC_{50} 3.026 mg ml^{-1}) followed by methanol (IC_{50} 4.278 mg ml^{-1}). Extracts obtained using water also showed equivalent scavenging activity (IC_{50} 4.762 mg ml^{-1}). These values were lower than those obtained using BHT (IC_{50} 2.830 mg ml^{-1}), but the IC_{50} values of the methanol and water extracts were comparable with α-tocopherol (IC_{50} 4.546 mg ml^{-1}). The extracts of *K. alvarezii* showed better radical scavenging activity than the extract of *P. palmate* (dulse) IC_{50} – 12.5 mg ml^{-1} (Yuan, Carrington, & Walsh, 2005a), and purified extract of *E. cava* IC_{50} – 5.49 x 10^3 μg.ml^{-1} (Suja, Jayalekshmy & Arumughan, 2005). Ragan and Glombitza (1986) reported the radical scavenging activity of seaweeds to be mostly related to their phenolic contents. On the other hand, Siriwardhana, Lee, Kim, Ha, & Jeon, (2003) and Lu and Foo (2000) reported a high correlation between DPPH radical scavenging activities and total polyphenolics r = (0.971). In the present study the linear regression analysis of DPPH scavenging (i.e EC_{50} values) with the total phenol content (gallic acid equivalents) gave an r value of 0.937, showing statistically significant correlation. *K. alvarezii* is the main industrial source of carrageenans (having alternating D–galactose 4-sulphate and 3,6–anhydro D–galactose residues), which may also contribute to the antioxidant potential of this seaweed. Components such as low molecular weight polysaccharides, pigments, proteins or peptides also influence the antioxidant activity (Siriwardhana, Lee, Kim, Ha, & Jeon, 2003).

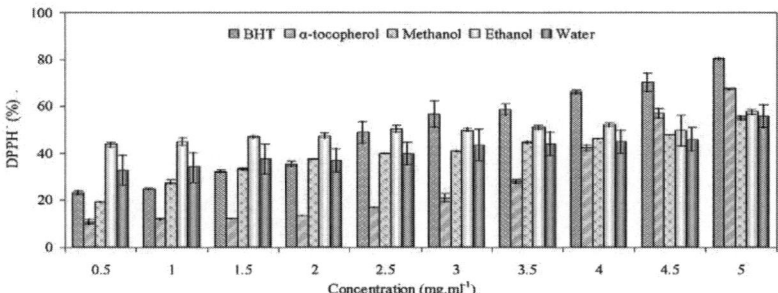

Fig. 1. Antioxidant activities of different solvent extracts of *K. alvarezii* determined as DPPH radical-scavenging activity.

3.4. Metal ion chelating activity

All the extracts demonstrated reasonable ferrous ion chelating efficacy (Fig. 2). The ascorbic acid extract demonstrated best ferrous chelating efficacy (IC_{50} 2.879 mg ml^{-1}) followed by methanol, ethanol and ethyl acetate (IC_{50} 3.075, 3.831 and 4.383 mg ml^{-1} respectively). Iron is known to generate free radicals through the Fenton & Haber–Weiss reaction. Metal ion chelating activity of an antioxidant molecule prevents oxyradical generation and the consequent oxidative damage. Metal ion chelating capacity plays a significant role in antioxidant mechanism since it reduces the concentration of the catalyzing transition metal in LPO. It is reported that chelating agents that form σ-bonds with a metal, are effective as secondary antioxidants since they reduce the redox potential thereby stabilizing the oxidized form of the metal ion (Srivastava, Harish, & Shivanandappa, 2006). Metal binding capacities of dietary fibers are well known the inhibitory effects on ferrous absorption of algal dietary fibers such as carrageenan, agar and alginate, were reported (Harmuth-Hoene & Schelenz, 1980). In this present study the carrageenan might have caused the decrease of ferrous ion in the assay system.

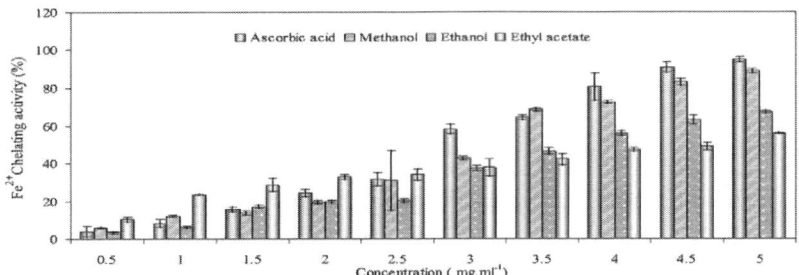

Fig. 2. Ferrous ion-chelating activities of different solvent extracts of *K. alvarezii*

3.5. Measurement of reducing potential

The reducing power of *K. alvarezii* extracts was concentration dependent (Fig. 3). As the concentration increased from 0.5 to 5.0mg ml^{-1}, there was an increase in absorbance with all the solvents except hexane. However the reducing power of the samples were found in the following order: BHT (0.23 – 0.879) > Methanol (0.07 – 0.74) > Ethanol (0.333 – 0.44) >Ethyl acetate (0.013 – 0.467) > Water (0.017 – 0.193) > Hexane (0.017 – 0.16). It is believed that antioxidant activity and reducing power are related. Reductones inhibit LPO by donating a hydrogen atom and thereby terminating the free radical chain reaction (Srivastava, Harish, & Shivanandappa, 2006).

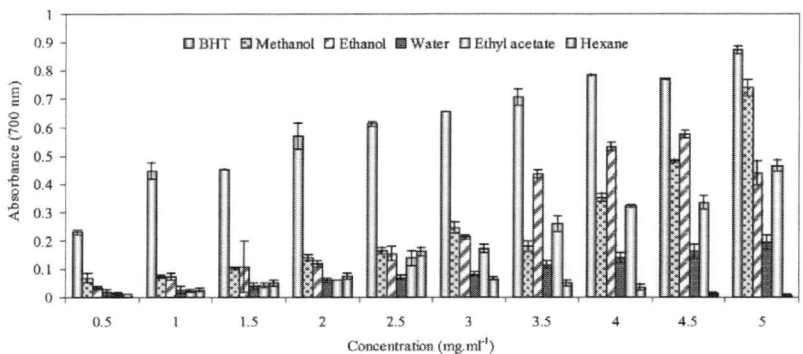

Fig. 3. Reducing powers of *K. alvarezii* extracts, along with a synthetic antioxidant

84

3.6. Antioxidant activity in linoleic acid system with ferrothiocyanate reagent (FTC)

Peroxyl radicals are formed by a direct reaction of oxygen with alkyl radicals. Decomposition of alkyl peroxides also results in peroxyl radicals. Peroxyl radicals are good Oxidizing agents, having more than 1000 mV of standard reduction potential (Decker, 1998). They can abstract hydrogen from other molecules with lower standard reduction potential. This reaction is frequently observed in the propagation stage of lipid peroxidation. Cell membranes are phospholipid bilayers with extrinsic proteins and are the direct target of lipid oxidation (Girotti, 1998). As lipid oxidation of cell membranes increases, the polarity of lipidphase surface charge and formation of protein oligomers increase; and molecular mobility of lipids, number of SH groups, and resistance to thermal denaturation decrease. Malonaldehyde, one of the lipid oxidation products, can react with free amino group of proteins, phospholipid, and nucleic acids leading to structural modification, which induce dysfunction of immune systems. The antioxidant effects of *K. alvarezii* extract and BHT on the peroxidation of linoleic acid were investigated and the results are presented in (Fig. 4). The decreasing absorbance indicates the increasing activity. The absorbance range recorded for control, BHT and sample were 0.0087–0.0151, 0.0021–0.0093 and 0.0037–0.0104 respectively. The ethanolic extract of *K. alvarezii* showed higher inhibitory effect than positive control BHT. This might be due to the presence of ascorbic acid and vitamin A (β–carotene) content in the extract of *K.alvarezi* (Fayaz, Namitha, Chidambara Murthy, Mahadeva Swamy, Sarada, Salma Khanam, SubbaRao & Ravishankar, 2005).

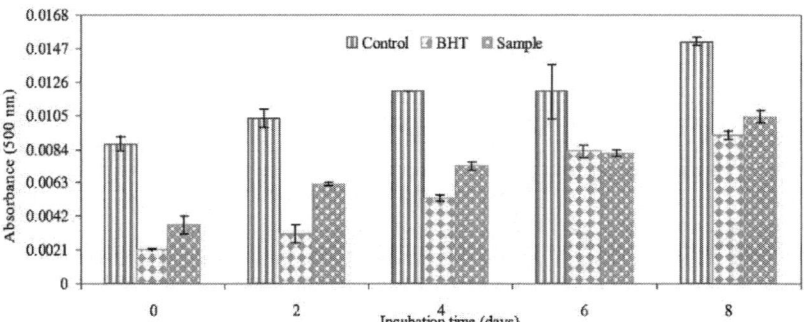

Fig. 4. Inhibitory effect of *K. alvarezii* extract on the primary oxidation of a linoleic acid system, using the ferric thiocyanate method.

Algal polysaccharides play an important role as free-radical scavengers in vitro and antioxidants for the prevention of oxidative damage in living organisms. Their activity depends on several structural parameters such as the degree of sulfation (DS), the molecular weight, the sulfation position, type of sugar and glycosidic branching. Moreover, some reports reveal that the sulfate and phosphate groups in the polysaccharides lead to differences in their biological activities. In vitro antioxidant activity of κ-carrageenan oligosaccharides and their oversulfated, acetylated, and phosphorylated derivatives have been reported by Yuan, Zhang, Li, Lu, Li, Gao & Songa (2005b). He also reported that phosphorylated and sulfated glucans exhibited better antioxidant ability than glucans and other neutral polysaccharides, which indicate that polyelectrolytes, such as glucan sulfate or phosphate, might have increased scavenging activity. Moreover, the sulfate content of polysaccharides from *P. yezoensis* was reported to contribute to the antioxidant activity. The cell wall of *K. alvarezii* is known to constitute of carrageenan, a sulfated polysaccharide, which may contribute to its antioxidant potential in addition to presence of ascorbic acid, vitamin A and various phenolic contents.

4. Conclusion

In the present investigation the various solvent extracts of *K. alvarezii* exhibited excellent scavenging effect (%) on DPPH, reducing power, ferrous ion chelating activity and antioxidant property on linoleic acid system. Thus it could be used in nutraceutical and in functional food applications. Since this is a preliminary study, a detailed investigation on composition of each component involved is absolutely necessary to authenticate the proper application which may open new frontiers for the human consumption of this seaweed world wide.

References

Blois, M.S. (1958). Antioxidant determinations by the use of a stable free radical. *Nature, 26,* 1199–1200.

Brand-Williams, W., Cuvelier, M.E., & Benset, C. (1995). Use of free radical method to evaluate antioxidant activity. *Lebensm Wiss Technol, 28,* 25–30.

Chapman, V.J., & Chapman, D.J. (1976). The Algae, ELBS & Macmillan: Basingstoke UK, 1–497.

Decker, E.A. (1998). Antioxidant mechanisms. In C. C. Akoh & D. B. Min (Eds.), *Food lipids, chemistry, nutrition, and biotechnology*, 397–401, New York.

Dinis, T.C.P., Madeira, V.M.C., & Almeida, L.M. (1994). Action of phenolic derivates (acetoaminophen, salicylate, and 5-aminosalicilate) as inhibitors of membrane lipid peroxidation and as peroxyl radical scavengers. *Archive Biochemistry Biophysics, 315,* 161–169.

Fayaz, M., Namitha, K.K., Chidambara Murthy, K.N., Mahadeva Swamy, M., Sarada, R., Salma Khanam, SubbaRao, P.V. & Ravishankar, G.A. (2005). Chemical composition, iron bioavailability, and antioxidant activity of *Kappaphycus alvarezii* (Doty) Doty, *Journal of Agricultural and Food Chemistry, 53,* 792–797.

Frlich, I., & Riederer, P. (1995). Free radical mechanisms in dementia ofAlzheimer type and the potential for antioxidative treatment. *Drug Research, 45,* 443–49.

Girotti, A. (1998). Lipid hydroperoxide generation, turnover, and effector action in biological systems. *Journal of Lipid Research, 39,* 1529–1542.

Huang H.-L., & Wang B.-G. (2004). Antioxidant Capacity and Lipophilic Content of Seaweeds Collected from the Qingdao Coastline. *Journal of Agriculture Food Chemistry, 52,* 4993–4997.

Harmuth-Hoene, A.E., & Schelenz, R. (1980). Effect on diatary fiber on mineral absorption in growing rats. *Journal of Nutrition, 110,* 1774–1784.

Jimenez-Escrig, A., Jimenez-Jimenez, I., Pulido, R., & Saura-Calixto, F. (2001). Antioxidant activity of fresh and processed edible seaweeds. *Journal of the Science of Food and Agriculture, 81,* 530–534.

Lu, Y., & Foo, Y.L. (2000). Antioxidant and free radical scavenging activities of selected medicinal herbs. *Journal of Life Science, 66,* 725–735.

Marja, P., Kähkönen., Anu., I. Hopia., Heikki, J. Vuorela., Rauha, J.P., Pihlaja, K., Kujala, T.S & Heinonen, M.(1999). Antioxidant Activity of Plant Extracts Containing Phenolic Compounds. *Journal of Agriculture Food Chemistry, 47,* 3954–3962.

Osawa, T., & Namaki, M. (1983). A novel type antioxidant isolated from leaf wax of *Eucalyptus* leaves. *Agricultural Biological Chemistry, 45,* 735–9.

Oyaizu, M. (1986). Studies on product of browning reaction prepared from glucose amine. *Japan Journal of Nutrition, 44,* 307–315.

Ragan, M.A. & Glombitza K.W. (1986). Phlorotannins, brown algal polyphenols. *Progress in Phycological Research, 4,* 129–241.

Rice-Evans, C., Miller, N.J., Bolwell, G.P., Bramley, P.M., & Pridham, J.B. (1995). The relative antioxidants activities of plant-derived polyphenolic flavonoids. *Free Radical Research, 22,* 375–383.

Singh, G., Maurya, S., de Lampasona, M.P., & Catalan, C. (2006). Chemical constituents, antifungal and antioxidative potential of *Foeniculum vulgare* volatile oil and its acetone extract. *Food Control, 17,* 745–752.

Singleton, V.L., & Rossi, J.A. (1965). Colorimetry of total phenolics with phosphomolybdic-phosphotungstic acid reagents. *American Journal of Enology and Viticulture, 16,* 144–158.

Siriwardhana, N., Lee, K.W., Kim, S.H., Ha, J.W., & Jeon, Y.J. (2003). Antioxidant activity of *Hizikia fusiformis* on reactive oxygen species scavenging and lipid peroxidation inhibition. *Food Science Technology International, 9,* 339–347.

Srivastava, A., Harish, S.R., & Shivanandappa, T. (2006). Antioxidant activity of the roots of *Decalepis hamiltonii* (Wight & Arn.). *LWT 39,* 1059–1065.

Sugihara, N., Arakawa, T., Ohnishi, M., & Furuno, K. (1999). Anti and pro-oxidative effects of flavonoids on metal induced lipid hydroperoxide-dependent lipid peroxidation in cultured hepatocytes located with ∝-linolenic acid. *Free Radical Biology and Medicine, 27,* 1313–1323.

Suja, K.P., Jayalekshmy, A., & Arumughan, C. (2005). Antioxidant activity of sesame cake extract. *Food Chemistry, 91,* 213–219.

Yan, X., Nagata, T., & Fan, X. (1998). Antioxidative activities in some common seaweeds. *Plant Foods for Human Nutrition, 52,* 253–262

Yuan, Y.V., Carrington, M.F., & Walsh, N.A. (2005a). Extracts from dulse (*Palmaria palmata*) are effective antioxidants and inhibitors of cell proliferation in vitro. *Food Chemical Toxicology, 43,* 1073–1081.

Yuan H., Zhang W., Li X., Lu X., Li N., Gao X. & Songa J. (2005b). Preparation and in vitro antioxidant activity of κ-carrageenan oligosaccharides and their oversulfated, acetylated, and phosphorylated derivatives. *Carbohydrate Research, 340,* 685–692.

Wilson D. (2000) Rhodophyta. Red Algae. Version 24 March 2000, http:// tolweb.org/ Rhodophyta/ 2381/2000.03.24.

Chapter V
Phycoremediation of heavy metals

1. Introduction

Heavy metal pollution is an environmental problem of worldwide concern. Some industrial processes result in the release of heavy metals in the natural water systems leading to increasing concern about their toxic effect as environmental contaminants. Lead (Pb), copper (Cu), cadmium (Cd), zinc (Zn) and nickel (Ni) are among the most common pollutants found in industrial effluents. Even at low concentrations these metals can be toxic to living organisms including humans. Metal sorption involves binding of metals onto the cell surface and to intracellular ligands (Mehta & Gaur, 2005). Conventional methods for removing heavy metals (precipitation, chemical oxidation/reduction, ion-exchange, reverse-osmosis, membrane separation, etc.) are often ineffective and costly (Marques et al., 1991; Volesky, 1994; Dey et. al., 1995; Kapoor & Viraragharvan, 1995; Lu & Wilkins 1996; Zhao & Duncan, 1997). Hence, biotechnological approaches have received a great deal of attention as an alternative tool in the recent years. Biosorption, the process of passive cation binding by dead or living biomass, provides a potentially cost-effective way of removing toxic metals from industrial wastewaters (Kuyucak & Volesky, 1990), and it could be employed most effectively in a concentration range below 100 mg L^{-1}, where other techniques are ineffective or costly (Vieira & Volesky, 2000; Schiewer & Volesky, 1995). Metal ion binding during biosorption processes has been found to involve complex mechanism, such as ion–exchange, complexation, electrostatic attraction and microprecipitation (Volesky & Holan, 1995). There have been some indications that ion-exchange plays an important role in metal sorption by algal biomass (Volesky et al., 2000).

Applicability of growing bacterial/fungal/algal cells for metal removal and the efforts directed towards cell/process development to make this option technically and economically viable for the treatment of metal rich effluents have been reviewed by Malik (2004). Recently biological removal processes have been attracting considerable attention for removing heavy metals from aqueous wastes. This includes new approaches like use of marine algal biomass for biosorption of heavy metals (Fourest & Roux, 1992; Mattuschka & Straube, 1993; Volesky, 1994; Volesky & Holan, 1995). Algae are of special interest in the search for and development of new biosorbents materials due to their high sorption uptake and their ready availability in practically unlimited quantities in the seas and oceans (Feng & Aldrich, 2004).

Marine algae (seaweeds) are readily available in large quantities for the development of highly effective biosorbent materials. However, considering the large number of macro-algal species identified so far, only a few have been studied for their heavy metal uptake properties. Most of these studies are limited to the *Ascophyllum* and *Sargassum* species (Yu, Matheickal, & Kaewsarn, 1999). The non–living biomass of marine algae, species of *Sargassum, Padina, Ulva,* and *Gracillaria*, have been investigated for their biosorption performance in the removal of lead, copper, cadmium, zinc, and nickel from dilute aqueous solutions. It was also found that the biosorption capacities were significantly affected by solution pH, with higher pH favoring higher metal-ion removal (Sheng et al., 2004). A number of workers investigated their feasibility of using cheaply available marine or fresh water algae for heavy metal removal (Darnall et al., 1986; Holan et al., 1993).

The passive removal of toxic heavy metals by brown marine algae via biosorption was reported by Davis et al., (2003a & b), who attributed this property to cell wall polysaccharides like alginate and fucoidan. Biosorption properties of a few algae are accredited to their cell-wall polysaccharides like alginate and fucoidan, which have a high affinity for divalent cations (Fourest et al., 1994; Puranik et al., 1999; Khoo & Ting, 2001; Chen et al., 2002; Davis et al., 2003b). The non-living biomass of *Sargassum* species, *Macrocystis pyrifera, Kjellmaniella crassiforia, Undaria pinnatifida* and *Ulva* species are known to effectively remove Cd, Cu, Zn, Cr and Ni (Seki & Suzuki, 1988; Yang & Volesky, 1999; Davis et al., 2000; Suzuki et al., 2005).

On the other hand, high adsorption capacities of various low cost adsorbents, e.g. chitosan (815, 273, 250, 222, 75 mg g^{-1} of Hg^{2+}, Cr^{6+}, Cd^{2+}, Cu^{2+}, and Zn^{2+}, respectively), zeolites (175 and 137 mg g^{-1} of Pb^{2+} and Cd^{2+}, respectively), waste slurry (1030, 560, 640 mg g^{-1} of Pb^{2+}, Hg^{2+}, and Cr^{6+}, respectively), and lignin (1865 and 95 mg g^{-1}of Pb^{2+} and Zn^{2+}, respectively) have been reported (Babel & Kurniawan, 2003). Removal of zinc from aqueous solutions using bagasse fly ash, waste from sugar cane industry as a low cost adsorbent has also been studied by Gupta & Sharma (2003). Fresh algal biomass of *Spirogyra* species was used as biosorbent for the removal of Cr (VI) from aqueous solutions (Gupta et al., 2001).

Kappaphycus alvarezii, a potential carageenophyte, is reported to occur in brown, green, pale yellow and red color forms in field cultivation (Ask & Azanza, 2002; Dawes, 1992; Hurtado-Ponce, 1995, Suresh Kumar et al., 2007a & b). The present study was carried out using three-color forms (brown, green and pale yellow) of *K. alvarezii* obtained from the cultivation farm of this seaweed at Port Okha, Northwest coast of India. The living biomass of these color forms was used for biosorption of Cd, Co and Cr and non-living biomass of the same forms was used for chelation of Cd, Co, Cr and Cu from aqueous solution in the laboratory. The results obtained are presented in this chapter.

a) Using live biomass of *Kappaphycus alvarezii*

2. Materials and methods

Heavy metal biosorption was studied using living biomass of *Kappaphycus alvarezii.*

2.1. Collection of samples

The three color forms - brown, green and pale yellow of *Kappphycus alvarezii* (Doty) Doty (Fig.1.) were obtained in April 2006 from the cultivation farm of Port Okha (L 22°28.528' N; L 069° 04.322' E), Northwest coast of India. The brown color form measured approximately 20-25 cm in size, was profusely branched with tapering tips and a thick basal part. The green color form also possessed a thick branching structure with tender tips; their size was smaller than the brown counterparts i.e. approximately 20 cm. Contrarily, the pale yellow form was smaller than the other two color forms, possessed less branches; the branches were very tender and broke even under moderate water current; the plant measured up to 14 – 16 cm.

(a)   (b)   (c)

Fig. 1. Three color forms of *K. alvarezii* (a) Brown, (b) Green and (c) Pale Yellow

The collected fresh algae were thoroughly washed with sterilized seawater to eliminate the adhering foreign materials, such as sand and debris, and materials were used for heavy metal uptake studies.

2.2. Heavy metal sorption

The following sets of experiments for biosorption studies were designed. Three 500 ml flat bottom flask with spout for air circulation were used containing 400 ml of sterilized seawater to which 25 mg L^{-1} each of cadmium, cobalt and chromium, as cadmium sulfate, cobalt sulfate and potassium dichromate (analytical grade) was added (pH 7.7). To avoid heavy metal contamination, the glassware was soaked in 10% HNO_3 for 24 h, rinsed with deionised water and oven dried prior to use.

Ten grams of each living algal color form (a single thallus fragment) was added aseptically to the three respective flasks containing mixture of heavy metals. One set without heavy metal was also inoculated with each color form that served as control. Experiments were conducted in duplicate. The living color forms were maintained at 25 ± 1 °C, in a clean environment where aeration was continuously provided. After 5 days of incubation, the algae were removed and shade dried at room temperature followed by drying in an oven at 80 °C for 1 h in a porcelain crucible. These samples were ashed at 550 °C in muffle furnace for 2 h. The ash was cooled at room temperature, moistened with 10 drops of distilled water and carefully dissolved in 3 ml HNO_3 (1:1 v/v). The crucible was then heated on a hot plate at 110 °C till the acid solution totally evaporated.

The crucible was returned to muffle furnace and ashed again for 1 h at 550 °C and cooled. Subsequently the ash was dissolved in 10 ml of HCl (1:1 v/v), and the solution was filtered through millipore filter paper (0.25 μ) into a 50 ml volumetric flask and 2 ml 0.1N HNO_3 was added to the filtrate and the final volume was made up to 50 ml using distilled water (Jones, 1984). This was subjected to heavy metal analysis.

Analysis of heavy metals cadmium (Cd), cobalt (Co) and chromim (Cr) was carried out using Inductively Coupled Plasma Optical Emission spectroscopy, ICP-OES (Perkin-Elmer, Optima 2000). The mean value of the results obtained here was considered and the heavy metal uptake was calculated as mg 100 g f.wt^{-1} (100 g of fresh algae approximately yielded 10 g of dry material).

3. Results and discussion

The three-color forms of *Kappaphycus alvarezii* were compared for their metal biosorption efficiency, as well as survival under stressed condition (i.e. presence of different heavy metal) after 5 days of incubation. It was observed that all the three color forms could survive in this stress environment. This could be considered as a positive indication for this seaweed to be used for phycoremediation.

Terry & Stone (2002) reported that living algae are known to adsorb more heavy metals due to metabolic uptake and continuous growth, e.g. cadmium and copper biosorption by *Scenedesmus abundans*. Sloof et al., (1995) found that cadmium uptake by living *Selenastrum capricornutum* was rapid in the first adsorption stage, and then continued more slowly in the physiological metabolic stage. Resistance, accumulation and allocation of zinc in two ecotypes of the green alga *Stigeoclonium tenue* Kütz coming from different habitats with different heavy metal concentrations has been reported (Pawlik-Skowronska, 2003).

As seen in Table 1, heavy metal treated brown color form of *Kappaphycus alvarezii* could sorb maximum metals, i.e. cadmium, cobalt and chromium 3.064, 3.365 and 2.799 mg 100 g FW (i.e. fresh weight) $^{-1}$ respectively. The pale yellow color form treated with metal could take up only 0.961, 1.403 and 0.942 mg 100 g FW^{-1} of cadmium, cobalt and chromium respectively.

Table 1

Biosorption (mg/100 g f.wt.) of heavy metals by three-color forms of *Kappa-phycus alvarezii*

Color form	Metal	Control	Biosorption
Brown	Cd	0.0225	3.064
	Co	0.0060	3.365
	Cr	0.0040	2.799
Green	Cd	0.0050	2.684
	Co	0.0040	3.430
	Cr	0.0045	2.692
Pale yellow	Cd	0.0310	0.961
	Co	0.0160	1.403
	Cr	0.0230	0.942

Similarly, the green color form showed a biosorption of 2.684, 3.430 and 2.692 mg 100 g FW $^{-1}$ of cadmium, cobalt and chromium respectively. However, the brown color form is the superior one in biosorption of cadmium and cobalt followed by green and pale yellow color forms of this seaweed. Biosorption of cobalt in the brown color form is slightly less than the green one. The control contained very little amount of heavy metals reflecting that these are absorbed from the surrounding environment (seawater) where they are cultivated. It is observed that all the three color forms of *Kappaphycus alvarezii* exhibited high cadmium, cobalt and chromium adsorption capacities when tested as fresh material. This is of extreme significance, as most algae do not survive in stressed environment containing heavy metals. The algae could not only survive up to a period of 5 days but also retained its unique ability to remove heavy metal ions. High adsorption efficiency of the algae, low biomass cost (mainly transportation cost), less labor input and high yields of biomass under cultivation makes this process of biosorption an effective, cheap and alternative technique for treatment of metal-bearing polluted marine environment. Employing this seaweed for biosorption studies fulfils the parameters like ecofriendliness and economic feasibility as suggested by Mehta & Gaur (2005).

Biosorption of cadmium and copper contaminated water by *Scenedesmus abundans* revealed that living algae could reduce cadmium from 10 to 0.10 mg.L^{-1} in 36 h (Terry & Stone, 2002). In the present study the brown color form of *Kappaphycus alvarezii* proved competent enough as it could adsorb 3.064 mg cadmium. 100 g FW^{-1}. Biosorption efficiency of brown algae, *Macrocystis pyrifera*, *Kjellmaniella crassiforia* and *Undaria pinnatifida* have been exploited for the recovery of lead and cadmium ions (Seki & Akira, 1998). It has been reported that alkali-pretreated *Ulva* biomass showed the sorption capacity (qm) from 60 to 90 mg g^{-1} and the sorption affinity from 0.03 to 0.04 mg L^{-1} at pH 7.8 while studying the uptake of cadmium, copper and zinc (Suzuki et al., 2005).

The study conducted here showed that the living biomass of *K. alvarezii* can be used as a potential candidate for biosorption studies.

b) Using non-living (dried) biomass of *Kappaphycus alvarezii*

4. Materials and methods

Heavy metal chelation was studied using non-living biomass of *Kappaphycus alvarezii.*

4.1. Collection of samples

The three color forms of *Kappaphycus alvarezii* were collected as mentioned above. These samples were thoroughly washed with seawater to remove epiphytes and dirt particles and then shade dried for two days. They were brought to the laboratory and oven dried at 80 ^{0}C for 3 h to obtain a constant weight and pulverized in the grinder (size 2 mm). This nonliving biomass was used for further experiments.

4.2. Heavy metal sorption

Chelation of heavy metals was studied by adding one-gram sample of each color form to Erlenmeyer flasks containing each 100 ml aqueous solutions of Cd, Co, Cr and Cu at four concentrations i.e. 25, 50, 75 and 100 mg L^{-1} prepared using analytical grade cadmium sulfate, cobalt sulfate, potassium dichromate and copper sulfate. Controls for each color form were also maintained without addition of heavy metals. Initial the pH was adjusted to 4.5 and at the conclusion of experiment the pH was found to be 5.3. All the experiments were conducted in triplicate. To avoid heavy metal contamination, the glassware were soaked in 10% HNO$_3$ for 24 h and rinsed with deionized water prior to use. The flasks were incubated for 72 h on a shaker at room temperature (33 ± 1°C) and subsequently the contents in the flask were syringe filtered (0.22 μm pore size, Millipore India Pvt. Ltd., Bangalore) and subjected to further analysis.

Heavy metal content in all the filtrates were quantified using Inductively Coupled Plasma Optical Emission Spectroscopy- ICP-OES (Perkin Elmer Optima-2000 DV). The difference between the amount of metal present in the filtrate and the total amount of metal present initially in the flask yielded the amount of metal chelated by the biomass. The percentage chelation for each metal was estimated. Mean and standard deviation were finally calculated.

5. Results and discussion

The chelating efficiency of the non-living biomass of the three color forms of *K. alvarezii* (Fig. 1) revealed that maximum amount of Cd, Co and Cu were chelated (%) when the heavy metal concentration was 25 mg L^{-1} i.e. the brown chelated Cd -5.37± 0.59, Co- 21.19± 0.13, Cu- 59.53 ± 0.37; green chelated Cd - 8.87± 0.61, Co-28.54 ± 0.13, Cu-70.87 ± 0.50 and pale yellow chelated Cd -15.84 ± 0.32 , Co- 32.32 ± 0.62 , Cu- 90.28 ± 0.89. In case of Cr maximum chelation was observed at 100 mg L^{-1} concentration i.e. brown- 68.79 ±0.20, green-77.21 ±0.42 and pale yellow- 88.10 ±0.15.

The chelation decreased with increasing concentration of Cd, Co and Cu while the same increased with increasing concentration of Cr. Further, no chelation had been found for Cd and Co at 100 mg L^{-1} concentration in brown color form, while the same was recorded for Cd only in green color form. Among all the four metals used, highest chelation was recorded in Cr followed by Cu, Co and Cd in the brown and green forms. But a different trend was observed in the pale yellow form, where maximum chelation of Cu was noted, followed by Cr, Co and Cd. This color form chelated highest amount of heavy metals, followed by green and brown forms. The pale yellow form of *K. alvarezii* is known to chelate 65.28 ± 0.51 to 90.28 ± 0.89 of copper and 80.86 ± 0.86 to 88.10 ± 0.15 % chromium. Lee, Park, Yang, Jeong, & Rhee (2000) have demonstrated 57% chromate chelation by a red algae *Pachymeniopsis* species. One of the brown seaweed, *Padina* species is known to chelate copper (0.80 mmol g^{-1}) (Kaewsarn, 2002).

Dried biomass of *Sargassum wightii* is also reported to chelate maximum cadmium metal followed by lead, copper and zinc, indicating the affinity range for heavy metal ions (Kumar & Kaladharan, 2006). Brown seaweed biomass of *Ecklonia maxima, Macrocystis angustifolia* and *Laminaria pallida* are known to sequester copper, zinc and cadmium ions at concentrations likely to be encountered in waste water, 0 – 100 mg L^{-1} (Stirk & Staden, 2000). *Ascophyllum nodosum* out performed a commercial ion exchange resin DUOLITE GT-73 by accruing 100 mg Cd^{2+} g^{-1} biomass (Holan et al., 1993). Hashim & Chu (2004) and Tsui et al., (2006) observed that brown seaweeds exhibit better metal chelation properties than their red counterparts. In contrast, in the present study, *K. alvarezii* has demonstrated more chelation of heavy metals. In particular, a notable increase in chelation of chromium (68.80-88.08 %) was recorded by increasing the concentration of Cr in the aqueous solution. The present investigation shows that all the three color forms of *K. alvarezii* exhibit excellent heavy metal chelating capacity, thus proving to be a potential biodetoxifier or metal scavenger.

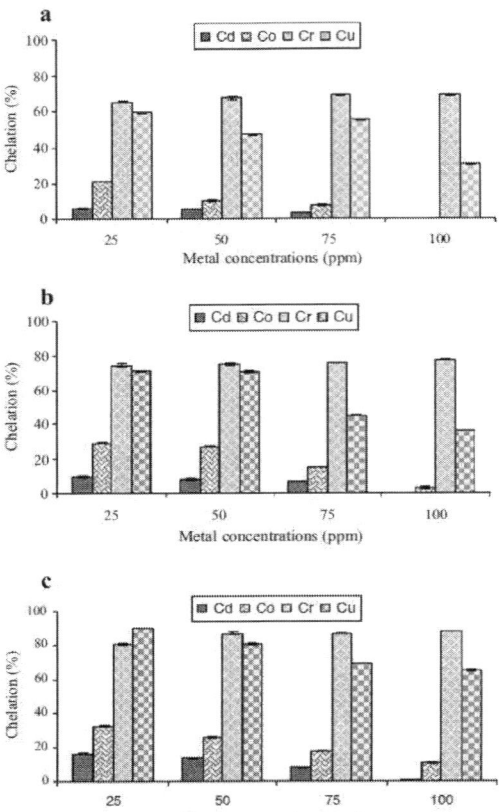

Fig. 1a–c Heavy metal chelation by three color forms of *Kappphycus alvarezii*. a Brown, b green, c pale yellow

The non-living biomass of all the three color forms of *K. alvarezii*, proved to be an efficient biosorbent, especially for chelation of Cr and Cu from aqueous solution. This is in contrast with the results obtained by Vieira & Volesky (2000), who opined that red marine algae containing carageenan do not have outstanding metal–sorbing properties though they have potential binding sites. The greater chelating capacity of the pale yellow form of *K. alvarezii* in comparison with that of other two forms (brown and green) might be attributed to the cumulative inherent physiology and the chemical composition (pigments, polysaccharide, etc.). Since the experiments conducted here are of preliminary nature, further studies need to be carried out to understand the mechanism and kinetics of heavy metal uptake which may broaden the scope of utilization of these three color forms for bioremediation.

6. Conclusion

Cultivation and use of this seaweed as a biodetoxifier in different parts of the world could be embarked upon as a greener (environment friendly) and profitable approach leading to employment generation for coastal living people on one hand and cleaning the environment on the other hand. This is an eco-friendly option for remediation of various coastal waters and would help develop a new scope of research.

References

Ask, E.I., & Azanza, R.V. (2002). Advances in cultivation technology of commercial eucheumatoid species: a review with suggestions for future research. *Aquaculture, 206*, 257–277.

Babel, S., & Kurniawan, T.A. (2003). Low-cost adsorbents for heavy metals uptake from contaminated water: a review. *Journal of Hazardous Materials, 97*, 219–243.

Chen, J.P., Hong, L., Wu. S.N., & Wang, L. (2002). Elucidation of interactions between metal analysis and modeling simulation. *Langmuir, 18*, 9413–9421.

Darnall, D.W., Greene, B., Hosea, M., Mcpherson, R.A., Henzl, M., & Alexander M.D. (1986). Recovery of heavy metals by immobilized algae. In: Thompson R (Ed.), Trace Metal Removal from Aqueous Solutions, The Royal Society of Chemistry, London, pp. 1–25.

Davis, T.A., Llanes, F., Volesky, B., & Mucci, A. (2003a). Metal selectivity of *Sargassum* spp. and their alginates in relation to their L-guluronic acid content and conformation. *Environmental Science and Technology, 37*, 261–267.

Davis, T.A., Volesky, B., & Alfonso, M. (2003b). A review of the biochemistry of heavy metal biosorption by brown algae. *Water Research, 37*, 4311–4330.

Davis, T.A., Volesky, B., & Vieira, R.H.S.F. (2000). *Sargassum* seaweed as biosorbent for heavy metals. *Water Research, 34*, 4270–4278.

Dawes, C.J. (1992). Irradiance acclimation of the cultured Philippines seaweeds, *Kappaphycus alvarezii* and *Eucheuma denticulatum*. *Botanica Marina, 35*,189–195.

Dey, S., Rao, P.R.N., Bhattacharyya, B.C., & Bandyopadhyay, M. (1995). Sorption of heavy metals by four basidiomycetous fungi. *Bioprocess Engineering, 12*, 273 –277.

Feng, D., & Aldrich, C. (2004). Adsorption of heavy metals by biomaterials derived from marine alga *Ecklonia maxima*. *Hydrometallurgy, 73*, 1–10.

Fourest, E., & Roux, J.C. (1992). Improvement of heavy metal biosorption by dead biomasses (*Rhizopus arrhizus, Mucor miehei* and *Penicillium chrysogenum*) pH control and cationic activation. *Applied Microbiology and Biotechnology, 37*, 399–403.

Fourest, E., Canal, C., & Roux, J.C. (1994). Improvement of heavy metal biosorption by dead biomasses (*Rhizopus arrhizus, Mucor miehei* and *Penicillium chrysogenum*) pH control and cationic activation. *FEMS Microbiology Reviews, 14*, 325–332.

Gupta, V.K., & Sharma, S. (2003). Removal of zinc from aqueous solutions using bagasse fly ash - a low cost adsorbent. *Industrial and Engineering Chemistry Research, 42*, 6619–6624.

Gupta, V.K., Srivastava, A.K., & Jain, N. (2001). Biosorption of chromium (VI) from aqueous solutions by green algae *spirogyra* species. *Water Research, 35*, 4079–4085.

Hashim, M.A., & Cheu, K.H. (2004). Biosorption of cadmium by brown, green and red seaweeds. *Chemical Engineering Journal, 97*, 249–255.

Holan, Z.R., Volesky, B., Prasetyo, I., & Stirk, W.A. (1993). Biosorption of Cadmium by biomass of marine algae. *Biotechnology and Bioengineering, 41*, 819 – 825.

Hurtado-Ponce, A.Q. (1995). Carrageenan properties and proximate composition of three morphotypes of *Kappaphycus alvarezii* Doty (Gigartinales, Rhodophyta) grown at two depths. *Botanica Marina, 38*, 215–219.

Jones, J.B. (1984). Plants. In: William S (Ed.) An Official Method of Analysis. Association Official Analytical Chemists, Arlington, VA, USA. pp. 38–64.

Kaewsarn, P. (2002). Biosorption of copper (II) from aqueous solutions by pre-treated biomass of marine algae *Padina* sp. *Chemosphere, 47*, 1081–1085.

Kapoor, A., & Viraragharvan, T. (1995). Fungal biosorption - an alternative treatment option for heavy metal bearing wastewaters: a review. *Bioresource Technology, 53*, 195–206.

Khoo, K.M., & Ting, Y.P. (2001). Biosorption of gold by immobilized fungal biomass, *Biochemical Engineering Journal, 8*, 51.

Kumar, V., & Kaladharan, P. (2006) Biosorption of metals from contaminated water-using seaweed. *Current Science, 90*, 1263–1267.

Kuyucak, N., & Volesky, B. (1990). Biosorption by algal biomass. In: Volesky B (Ed.), Biosorption of Heavy Metals . CRC Press, Boca Raton, Florida, USA. pp. 175.

Lee, D.C., Park, C.J., Yang, J.E., Jeong, Y.H., & Rhee, H.I. (2000). Screening of hexavalent chromium biosorbent for marine algae. *Applied Microbiology and Biotechnology, 54*, 445–448.

Lu, Y., & Wilkins, E.J. (1996). Heavy metal removal by caustic-treated yeast immobilized in alginate. *Journal of Hazardous Materials, 49*, 165–179.

Malik, A. (2004). Metal bioremediation through growing cells. *Environment International, 30*, 261–278.

Marques, A.M., Roca, X., Dolores Simon-Pujol, M., Carmen Fuste, M., & Congregado, F. (1991). Uranium accumulation by *Pseudomonas* sp. EPS-5028. *Applied Microbiology and Biotechnology, 30*, 406–410.

Mattuschka, B., & Straube, G. (1993). Biosorption of metals by a. waste biomass. *Journal of Chemical Technology and Biotechnology, 58*, 57–63.

Mehta, S.K., & Gaur, J.P. (2005). Use of algae for removing heavy metal ions from wastewater: progress and prospects. *Critical Reviews in Biotechnology, 25*, 113–152.

Pawlik-Skowrónska, B. (2003). Resistance accumulation and allocation of zinc in two ecotypes of the green alga *Stigeoclonium tenue Kutz* coming from habitats of different heavy metal concentrations. *Aquatic Botany, 75*, 189–198.

Puranik, P.R., Modak, J.M., & Paknikar, K.M. (1999). A comparative study on the mass transfer kinetics of metal biosorption by microbial biomass. *Hydrometallurgy*, *52*, 189–197.

Schiewer, S., & Volesky, B. (1995). Modelling of the proton-metal ion exchange in biosorption. *Environmental Science and Technology*, *29*, 3049–3058.

Seki, H., & Akira, S. (1998). Biosorption of heavy metal ions to brown algae, *Macrocystis pyrifera, Kjellmaniella crassiforia* and *Undaria pinnatifida. Journal of Colloid and Interface Science*, *206*, 297–301.

Seki, H., & Suzuki, A. (1998). Biosorption of heavy metal ions to brown algae, *Macrocystis pyrifera, Kjellmaniella crassiforia* and *Undaria pinnatifida. Journal of Colloid and Interface Science*, *206*,297–301

Sheng, P.X., Ting, Y.P., Chen, P.J., & Hong, L. (2004). Sorption of lead, copper, cadmium, zinc, and nickel by marine algal biomass: characterization of biosorptive capacity and investigation of mechanisms. *Journal of Colloid and Interface Science*, *275*, 131–141.

Sloof, J.E., Viragh, A., & Van Der Veer, B. (1995). Kinetics of cadmium uptake by green algae. *Water Air Soil Pollution*, *83*, 105–122.

Stirk, W.A., & Van Staden, J. (2000). Removal of heavy metals from solution using dried brown seaweed material, *Botanica Marina*, *43*: 467–473.

Suresh Kumar, K., Ganesan, K., & Subba Rao, P.V. (2007a). Phycoremediation of heavy metals by three- color forms of *Kappaphycus alvarezii. Journal of Hazardous Materials*, *143*, 590–592.

Suresh Kumar, K., Ganesan, K., & Subba Rao, P.V. (2007b). Heavy metal chelation by non-living biomass of three color forms of *Kappaphycus alvarezii* (Doty) Doty. *Journal of Applied Phycology,* DOI -10-1007/s10811-007-9181-8.

Suzuki, Y., Kametani, T., & Maruyama, T. (2005). Removal of heavy metals from aqueous solution by nonliving *Ulva* seaweed as biosorbent. *Water Research*, *39*, 1803–1808.

Terry, P.A., & Stone, W. (2002). Biosorption of cadmium and copper contaminated water by *Scenedesmus abundans. Chemosphere*, *47*, 249–255.

Tsui, M.T.K., Cheung, K.C., Tam, N.F.Y., & Wong, M.H. (2006). A comparative study on metal sorption by seaweed. *Chemosphere*, *65*, 51–57.

Vieira, R.H., & Volesky, B. (2000). Biosorption: a solution to pollution. *International Microbiology*, *3*, 17–24.

Volesky, B. (1994). Advances in biosorption of metals: selection of biomass types. *FEMS Microbiology Reviews*, *14*, 291–302.

Volesky, B., Figueira, M.M., Ciminelli, V.S., & Roddick, F.A. (2000). Biosorption of metals in brown seaweed biomass. *Water Research*, *34*, 196–204.

Volesky, B., & Holan, Z.R. (1995). Biosorption of heavy metals. *Biotechnology Progress*, *11*, 235– 250.

Yang, J., & Volesky, B. (1999). Biosorption of uranium on *Sargassum* biomass. *Water Research*, *33*, 3357–3363.

Yu, M.Q., Matheickal, J.T., Yin, P., & Kaewsarn, P. (1999) Heavy metal uptake capacities of common marine macroalgal biomass. *Water Research*, *33*, 1534–1537.

Zhao, M., & Duncan, J.R. (1997) Batch removal of sexivalent chromium by *Azolla filiculoides. Biotechnology and Applied Biochemistry*, *26*, 179–182.

Druck: KN Digital Printforce GmbH · Schockenriedstraße 37 · 70565 Stuttgart